The Joy of Learning to Fly

The Joy of Learning to Fly

GAY DALBY MAHER

AN ELEANOR FRIEDE BOOK

MACMILLAN PUBLISHING COMPANY
NEW YORK

COLLIER MACMILLAN PUBLISHERS
LONDON

Macmillan Publishing Company
866 Third Avenue, New York, N.Y. 10022
Collier Macmillan Canada, Inc.

Designed by MaryJane DiMassi

Library of Congress Cataloging in Publication Data

Maher, Gay Dalby.
The joy of learning to fly.

Bibliography: p.
Includes index.
1. Airplanes—Piloting. I. Title.
TL710.M355 629.132'52 78-4822
ISBN 0-02-579320-9

First Macmillan Edition 1984

10 9 8 7 6 5 4 3 2 1

Printed in the United States of America

ACKNOWLEDGMENTS

This book is dedicated to my students. They have been often confused, frustrated, and discouraged as I have slowly learned how to help. I owe to their patience and persistence most of what I know about teaching, much of what I know about flying, and a lot of what I know about being a person. I am deeply grateful.

More particular thanks are due to Theodore Maher, who made it all possible by babysitting at least three hours for every one of the forty hours I flew for my private license, and to Mark McCarty and Mary Louise DeSimone who encouraged me when I might otherwise have abandoned the effort to write this book.

Contents

Introduction

Pilots are people who know they're into something that's bigger than they are. Being part of the world of the air stretches us and makes us bigger than we would otherwise be. We are given constant reminders that the earth and sky are huge and untamed—reminders that most groundlings miss—and they keep us humble. But we are humble in a curiously proud way. The feeling is only one short step from a religious sense of being among the chosen. We bow to no man, but we know our master, and we are proud to acknowledge the awesome power and immensity of the natural world.

There are pilots who fly only to get somewhere, or to earn a good living. But most of us are deeply committed to all that being a pilot means. A popular psychological gimmick is to ask a person to answer the question "Who are you?" in three different ways. Most pilots would include, "I am a pilot," among their answers.

People begin to fly for many different reasons. Perhaps you're interested because your best friend is a pilot, and he's urging you to try it. Or maybe you fly cross-country with your boss in his light plane on business trips, and you've been thinking about what you'd do if he had a heart attack in the air. Maybe your spouse is taking lessons. And I don't use the word "spouse" just to please the feminists. There are many women pilots whose husbands don't fly. Let me make a statement here about those irritating personal pronouns. I know that women fly and that women teach. But it is simply too awkward to use he/she or "him or her," so I use the male pronouns throughout and they should be taken to refer equally to males or females.

Some people get into flying as soon as they finally can afford it—it's the fulfillment of a lifelong dream. At the opposite extreme I knew one student who learned to fly because he was so scared of it. A lot of airline traveling was essential to his job, and the only way he could get himself onto the airliner was to get drunk first. He thought he could get over that fear if he learned to fly himself. It wasn't easy —for him or for his instructor—but he did it and it did work. Whatever the reason for learning, the fact of flying, in the air, alone—the fact of being a pilot—becomes important. For me it was the center of my life for many years. And it is only the confidence, the self-esteem, that I learned as a pilot that has allowed me to branch out, to grow in other ways, to develop and enjoy other capacities.

I'm writing this book because flying has meant so much to me. I'm writing it because I've met so many people who've been discouraged from flying by poor instruction. I'm writing it, perhaps most of all, because I've learned so much in the years of struggling to find ways to help my students. In a sense, I've learned at their expense, and I feel I owe them a debt that I can repay only by doing my best to hand that knowledge on to you.

The people I'm writing for, the people I have in mind when I say "you," are people who are thinking about learning to fly—and about learning to grow too. You are not quite satisfied with yourselves, with your lives, with the way you practice the art of being a person. I've been there, where you are now. I remember how it felt to be a beginner. I remember the uncertainty and frustration and I remember the joy there was in each step as I progressed. But I don't have to depend only on memory. With about 9000 hours in the air and fifty years of living, it's sometimes hard to persuade people that I'm not just pulling a modesty-becomes-me act when I say I'm still a student. But it's true; and I know, out of my own experience, about the kinds of things that make it easier or harder for me to learn in the air.

Obviously I didn't acquire 9000 hours in just a couple of years. I took my first lesson in a Piper Cub in 1957. In those days most flying was strictly a fair-weather, daylight affair, without radios or instruments. But things were changing and by the time I had 500 hours of instruction behind me most airplanes had electrical systems and gyro instruments and new pilots had to demonstrate proficiency in using the radio and in flying by reference to instruments alone.

I wore a lot of hats on my first job in aviation. In fact, I did everything except instruct: cleaned the office, answered the phone, ordered supplies, paid bills, ran the counter, made the deposits, balanced the books, and occasionally flew passenger flights or parachute jumpers. When I finally got the instructor's rating I just added instruction to my other responsibilities. That was before the days of women's lib, and I didn't notice I was being exploited until they hired three people to replace me when I left: a pilot, a secretary, and a bookkeeper.

My second job was a big step up. Here the manager did most of the office work. I grew in confidence and technique, both as a pilot and as an instructor. And I gained confidence too, simply as a person. I had to run ground school and I found the classes frightening to think about, but fun to do, and after a while I wasn't frightened by the idea of being in front of a group of people, talking about what I knew. The owners of the field found that they could get some free publicity by taking advantage of the fact that a woman instructor was a rarity, and I learned that a woman instructor who was also "mother of three" was irresistible to the news media. I learned to enjoy the radio and newspaper interviews.

All along I was learning from my students—from the interactions with the various kinds of people who were trying to learn to fly. I had one flight student who was also a professional in his own field, engineering. From my struggles to help him, I learned a teaching trick I've used ever since. The problem was that I couldn't persuade this very intelligent man not to stare at the instruments during his approaches, and they were bad approaches—airspeed and nose attitude oscillating like a roller coaster. He knew the theory of flight as well as I did, but he couldn't translate his knowledge into practice. A product of a long professional training in his field, he had never learned to be aware of what his senses could tell him, nor to trust what he sensed. He needed gauges to make him comfortable. In desperation I covered the airspeed indicator. I was astonished and delighted to find that his first approach without the indicator was the best he'd ever made. After a few more lessons he no longer needed any coaching about whether the airplane's nose was too high or too low. I uncovered the airspeed indicator, and soon after that he soloed.

Several weeks later I arrived at the airport on Saturday morning and found that he had been dispatched by a new counter person in an airplane whose airspeed indicator didn't work at all—an airplane that was supposed to be grounded until the problem was corrected. My heart sank and I had terrifying visions of disaster. I remembered the trouble I'd had getting an airplane down once when I'd lost the airspeed indicator. And I hadn't been a new student either. So now I rushed outside and watched him making an approach. I expected to see wild gyrations and braced myself to watch him come to a stop in the bushes at the far end of the field. Instead his approach was smooth and the landing was perfect. I thought the problem with the indicator must have cleared itself up. When he came in after a few more landings I asked if the airspeed indicator had given him any trouble. He said he hadn't noticed anything wrong with it. I found it was still inoperative, so apparently he had never looked at it. Of course, he should have checked it, but for this particular pilot not looking at it was a victory. He had grown as a person in developing that new confidence in his own senses.

My next move was to Flying W Ranch in New Jersey in 1962. The years there could make a book in themselves. Many of aviation's greats were regular visitors. There couldn't have been a better place to develop as an instructor. Bill Whitesell, the boss, was a pilot himself, deeply committed to aviation, always willing to discuss points of technique or explain a prop-feathering system, or listen to a suggestion about the operation. A lot of experienced pilots were always around to learn from and hash things out with. And there was a lot of flying, not all of it instruction, to be done.

During the seven years I was there I made two record-breaking cross-country helicopter flights, flew two Powder Puff Derbies, flew an International Air Race from Nicaragua to Florida. I earned the multi-engine rating, the helicopter rating, the instrument and instrument instructor ratings, and all the ground instructor ratings. I soloed my oldest daughter there on her sixteenth birthday. Always I was learning and growing, and some learning and growing can be painful.

I developed one valuable teaching technique after I made

the most humiliating mistake of my career one busy weekend afternoon. I was flying with a student in her own Navion. She knew the airplane well and was letter-perfect in her procedures. She needed dual only because she had lost confidence after an accident she'd had in a strong crosswind. Today she was doing well in the crosswind in spite of the heavy weekend traffic, and we were both feeling pretty good about the way things were going. Then on downwind we spotted a twin making a long straight-in approach. We extended our downwind for him, and delayed putting the gear down. Maybe you can guess what happened. When I heard the horn blow, just off the ground after her beautiful crosswind approach, I thought it was the stall warning and was ready to congratulate her on a perfect landing. Then we crunched onto the runway, sparks flying, wheels still nicely tucked up into the wheel wells.

I knew it was entirely my responsibility and I expected to be fired, my career over. I didn't know there was a saying about landing with the gear up. It goes, "There are two kinds of pilots, those who have and those who will." Now I know it can happen even with a full professional crew, because I later saw a newspaper photo of a four-engine transport that had landed gear up at Dulles due to pilot error. When Bill called me into his office his first words were, "What did you learn?" Then he told me a trick that had saved him once in a transport plane in which there were a pilot *and* a copilot *and* a flight engineer. Bill's secret is to make a gear check at the last possible moment that a go-around can be initiated—just off the ground as you pull the wheel back or make the last power reduction.

Knowing about this method is a help, but learning to actually do it may not be so easy. To train people to do it I employ a kind of shock treatment which helps to develop a reflex response. The first time a student forgets this last check

I yell, "Go around!" when the plane is inches off the runway. He doesn't wait to ask why; he goes around. The shock of that moment makes a very strong impression. Most people have the last gear check habit ingrained after only one such experience.

Through all these years my personal life interacted with my professional life. I was divorced before I became an instructor, and the desire to be at home and available to my three children on a predictable basis led me to avoid charter and corporate flying. Then in 1965 I was offered complete control of the training operation, but the discovery that I had breast cancer and the subsequent operations and treatment took the decision out of my hands at that time. In 1967 new management took over the school at Flying W and I served as Chief Pilot for them for a year and a half. We were FAA (Federal Aviation Administration) approved and VA (Veterans Administration) approved and there were many things about the administration of such a school that I found challenging and fascinating. Dealing with officialdom was more frustrating than rewarding, but flying flight checks with other people's students in my capacity as Chief Pilot, and flight testing applicants for the private and commercial certificates as an FAA designated examiner were eye-opening. Most stimulating of all were the discussions and the flights with instructors who worked for me or who were applying for jobs.

But I found that I didn't really like being responsible for how other people did their jobs, especially when that responsibility ceased to be backed up by the authority to hire and fire as I saw fit. So in 1969 I resigned and began working for myself. I had a small office and flew in my own Cessna 172 or in the airplanes owned by students.

I now have thousands of hours in single-engine, fixed-gear airplanes with hundreds of students, and I have hundreds of hours with tens of students in more sophisticated aircraft. More than half of those hours were spent going around and around the pattern in daylight hours, in good weather. That's how most of the beginner's dual hours are spent. Then I have hundreds of hours of cross-country, both with and without radio; hundreds of hours giving instrument instruction, and more hundreds of hours giving night instruction. Out of my experience as an instructor I've developed some very strong opinions about the most effective ways to learn the fundamental skills of flying. And I have some strong feelings too about how good judgment can be encouraged. Skill and judgment together make flying all that it should be—safe, efficient, and full of joy. And both skill and judgment develop as the product of two things—the pilot's own basic nature and the instruction he receives.

The pilot must develop mature judgment and must be willing to take full responsibility for everything that happens to him. If the weather is worse than the weatherman said it would be, *it isn't the weatherman's fault,* nor is it bad luck. It's just one of those "unexpected" things the pilot must be ready for. Life isn't fair, and living isn't easy, though it seems to be part of human nature to keep trying to make it that way. But we mustn't get the idea, when life isn't fair and easy, that we are somehow relieved of the responsibility for dealing with it as it actually is.

My experience includes a lot of variety in both airplanes and airports. I've flown two-place tail-draggers, even biplanes, out of grass fields, and I've flown fully-equipped twin-engine craft out of huge international airports. I've watched myself make mistakes and I've watched other people make mistakes. Some have been funny, some have been fatal. Always I've tried to figure out how the mistake came to be made, and how instruction might be able to prevent its being made by others. And as my students and I have labored together to overcome confusion, tension, and discouragement, I've worked out methods of developing skill and judgment that seem to work for most people.

My instructional approach is rooted in my personal values and convictions. I believe above all that each individual is responsible for his own actions. I believe that the capacity for choice, the acceptance of responsibility, the internal honesty and strength of the fully responsible person are among the finest and most admirable qualities in human nature. These characteristics are fostered by flying. Flying constantly demands decisions of every pilot—some of them may be life-and-death decisions—and all these decisions require a deep-rooted sense of personal responsibility and a lot of self-discipline.

As my own boss I've been able to cancel a lesson and just talk when it seemed that that was what the student needed. You might say that the whole business has been oriented a little more toward people than toward aviation. And one overwhelming fact has emerged from this experience. Learning to fly, and getting a license, then going on and learning from every flight, helps people grow in many ways. It can make changes happen almost as therapy does. As people become really aware of the world their senses convey to them and of the challenges to mind and heart that the sky holds, they also discover new capacities within themselves to meet those challenges.

At first flying skills may serve only as a crutch to a limping ego, but that support allows a person's spirit to grow healthy and strong. There is a base from which he can take off as if jet-assisted—to grow as a person even as he continues to become more skillful as a pilot. Finally the confidence and self-esteem that are soundly based on competence in flying will carry over into everyday living and make his whole life richer and more rewarding.

Seeing this happen in my students and friends has been the most rewarding part of my career. Sometimes the routine of flying round and round the pattern, "circuits and bumps" the English call it, gets me down. But I love flying and I remember vividly what it felt like to be learning to fly. To be part of that joy for someone else can always excite and inspire me. Despite the crowded sky, I have a small wish that every human being could experience the glorious joy of flying an airplane alone.

I am heartsick when I talk to someone who has actually started to take lessons and then been frightened into quitting by the gyrations and showing-off of a macho instructor, or been discouraged by the lack of progress when the instructor was inept. Above all I wish that this book could help everyone who begins to learn to fly to find in it the same pleasure, the same challenge, and the same confidence that I have found.

—G.D.M.

The Joy of Learning to Fly

1. Personalities and Flying

Personalities—yours and your instructor's—and how they mesh will determine how fast and how well you learn to fly. How fast and how well you learn to fly will influence how you feel about yourself. Most students become deeply ego-involved in flying, and it can be important to have an instructor who understands that your self-esteem is at stake. While your own traits and aptitudes and how you already feel about yourself are basic, the relationship you develop with your instructor will have a great deal to do with how efficiently you can make use of the qualities you have.

Once you find a licensed instructor, the most important thing you need to know about him is whether he is interested in teaching *you* to fly. It may be nice to be able to brag about him: the 15,000 hours he spent bush-flying in Alaska, or the air shows he puts on, or the airplanes he designs and builds. But no matter how glamorous and impressive, none of that can compensate you if you are spending frustrating hours and wasted money trying to learn to fly with an instructor who doesn't really care about you and your progress.

Meeting the Instructor

There are a lot of ways an instructor can show you he cares without actually saying so. The first thing you'll notice is the atmosphere of the first interview. If he's in the middle of a busy flight office, with urgent phone calls and exotic, technical questions arising on every hand that only he can answer, you'll probably be impressed. But as you become more impressed with how important he is, you're also going to become impressed with how important you aren't. It will then be quite natural for you to become angry, withdrawn, or humbly grateful for the snatches of attention he allows you. None of these attitudes will help you learn to fly, so bid this busy fellow, "Good day." That's what an air traffic controller says when he means, "Good-bye. I don't expect to communicate with you again."

We'll assume you've found an instructor who does take the trouble to seek out a quiet place for your first interview, even if it's only the cockpit of the plane you're going to fly. Now ask yourself honestly if you're ready to like him. One of the standard discussions at FAA (Federal Aviation Administration) seminars centers on prejudices—superficial reasons for not liking people: long hair, pipe smoking, cowboy boots, female gender, dead-fish handshake. Unfortunately, even when we try to overcome the dislike or distrust we know is rooted in prejudice, a sense of uneasiness is likely to persist. If you are trying to take instruction, especially in the air, from someone whom you don't quite trust, you probably won't hear a lot of what he says. The result will be frustrating for both of you. You'll save a lot of your time and money if you avoid flying with anyone against whom you feel an unreasoning prejudice.

Looking at Instructors

Let's face it, most instructors would rather be flying than teaching flying. We learn to fly because we love flying, but we learn to instruct because it seems like the best way to fly and still continue to eat. There are only a few of us who, like me, always wanted to be instructors because we thought we'd learn more about flying if we were teaching. So it may be that you have more at stake in this relationship than your instructor has. And you may be able to contribute more to making the relationship work if you know something about instructors in general.

Most of us begin our careers by teaching the same way we were taught. We may leave out a few things our instructors did to us that we didn't like—hitting us over the head with a rolled-up chart or screaming at us, for example. And we may add some things we always wished we'd had—more demonstrations of landings as they should be done, or making a game of looking for traffic. And that's not a bad way to start out instructing.

But it takes many hours of experience before we begin to really do a good job of teaching to more than just a few of the varied students we work with. Not only is each person unique; each student-instructor combination is unique. The teaching method that works well in one combination may not work at all in another. Maybe I remember freezing when an instructor screamed at me, so that's a method I never use. But maybe the guy who's driving me crazy because he never seems to listen to me needs to be screamed at. Maybe he thinks that anything that isn't screamed at him just can't be very important. The decibel level I would normally use only when disaster is approaching, he reads as being just barely worth listening to. Until I figure all that out and begin to do some screaming, he's listening to me with only half an ear.

And even when I think I have a guy's measure and know

how to instruct him to fit his foibles, I'll often be wrong. Because there'll be days when he and I are different, and the combination is different, and I'll need to use a different approach. Maybe the lovely lady whose eyes have always filled with tears at the most gentle corrections needs, today, to be told plainly that it's time to stop being a crybaby. Maybe the guy who finally began to listen only when I began to punch his shoulder every time he did something wrong, needs some sympathy and understanding at this point. The fact is I can't be right all the time, and sometimes I'd like to just treat everyone the same and let them adjust to me. But I know I'll do a better job if I can be aware of my student's state of mind and respond to it. You can help your instructor if you tell him how you're feeling.

Working Hard Without Too Much Tension

If you feel tense and nervous, and especially if your instructor is making it worse, let him know. In learning to fly, you need to develop exactly the kind of physical awareness that's blocked out by tension. One thing you can do is make sure you don't fly when you aren't feeling really well. A hangover, a cold, or a headache are good reasons not to fly. An emotional upset may have a bad effect too. If you are able to leave the cares and problems of your life behind you on the ground, then go ahead and fly. But don't try to fly if you carry your troubles into the cockpit with you.

A cockpit is a kind of echo chamber for tension. A tense instructor makes most students more tense. A tense student makes most instructors more tense. Once it starts, this reverberating tension can become really dangerous. You may have heard a horror story about a student and instructor spinning in. It's not likely to happen in a modern airplane, but such a thing is possible if the student freezes on the controls at a critical moment. That's caused by nothing more mysterious than tension gone wild. So when you notice your muscles tightening and the instructor's voice getting a little loud or shrill, you can bet there's tension building in the cockpit. That's the time to slow down and ease off. You can tell the instructor you need to relax for a minute. Ask him if he'll take the controls, while you take a look around, or try to match the chart with landmarks, or tune the radio. Or he could show you some nice smooth maneuver, or go through the fundamentals as they should be done.

I used to think shrugging shoulders and wiggling toes was a little too gimmicky as a relaxation aid, but these little things do help. So shrug your shoulders and wiggle your toes before you take back the controls. And while you're relaxing, look around, listen to the engine and the sound of the air past the plane. Make yourself aware of what you're feeling in the seat of your pants. Of course, it may be that you won't notice that you're tense until your instructor suddenly says, loudly and firmly, "Relax, goddammit, relax!" Unfortunately, this instruction will probably tighten every one of your muscles to a point of cramp just this side of real pain. If tension is a major problem for you, you might do better with an instructor who is more understanding. But if you really like the instructor you have, you may be able to learn to be more aware of your own tension, and work on it yourself before he ever notices it. It's important to understand, though, that tension is a natural part of the process. Remember that what you and your instructor are both aiming at is to take a real live human being—you—and turn him loose in the sky, alone, with no one to help. You both want to be sure that you're going to be able to do the job.

Looking at Yourself

Of course, this student-instructor relationship is not a one-way thing. The instructor needs to like you too. Here you are: a male-female, old-young, shy-brash, intelligent-slow, agile-clumsy, pleasant-sarcastic, eager-timid person. You're about to spend a lot of money and time and put your ego on the line with this stranger. He needs to get to know you a little so he can best help you get a good start on the process that's eventually going to take you alone into the wild blue. Let him in on how you feel about flying. Tell him what your friends and relations think of the idea. If you already know something about flying, either from books or from experience in light planes, let him know about that too.

Tell him why you want to fly, and how you plan to use flying, so he'll know better how he can motivate you if you come to a time of discouragement. If you're flying for fun—because it seems like a neat thing to do, glamorous and exciting—you can be motivated by the instructor's attitude toward excellence. The beauty and grace of a really well-executed turn, the satisfaction of fitting efficiently into the flow of traffic, the pride in the perfect landing will be enough for you. You won't need the crowd of Sunday afternoon onlookers to make it count.

On the other hand, if your time is money; if you just don't have time to be doing anything just for the fun of it; if you're learning to fly because flying will save you time and make you money, then your instructor will use a different approach. Excellence in flying is efficient and safe. And safety and efficiency in flight will save you both time and money, making your airplane and your ability to fly it more valuable to you.

So there are these two major categories into which pilots can be divided: those who take to flight for its own sake, and those who learn to fly so they can travel faster. Not only do they have different reasons for wanting to fly, they also have different misconceptions about flying and learning to fly. Consider those who fall into the first category. Jill Dreamer thinks pilots are free, brave, wise, and never have colds. She and her brother, Joe Daring, also think instructors are kind and all-seeing. Actually all the attributes they think instructors have add up to the picture that most people have of God. So here is God, sitting right beside them telling them what to do so they can grow up to be junior gods themselves. They will certainly try to do what He tells them, so they are easy to teach.

If you are a Dreamer or a Daring you will have to learn, as you work at becoming a pilot yourself, that, in fact, pilots are people with approximately the same faults as everybody else. Pilots are free in ways that groundlings aren't, but we are subject to the laws of nature and we have to curb our freedom with a rein of good sense and self-discipline. We learn to accept our capacity to make mistakes, even while we strive to make fewer and fewer. The joys of flying are many, and even though becoming God is not one of them, we do sometimes experience a sense of exaltation.

The tycoon in the making is another story. Rick Moneymaker probably thinks that anybody who flies just for fun is wasting a lot of time and may be a little soft in the head too. Rick has places to go, and he wants to go in a hurry, and he figures he can learn to drive one of these things with wings, and then he can get there quicker. To him the instructor is part of a tiresome formality which keeps him from getting right in and driving through the sky to greener pastures. The instructor is an obstacle, and most of Rick's energy, which is considerable, will be devoted to finding ways to get around the requirements the instructor is trying to help him meet.

If you must view the airplane as a means of transportation

and only that, you'll miss most of the joy of flying. But even without that joy as a reward, you can't afford not to be a good pilot. In fact, if you plan to use the airplane regularly to keep appointments and meet deadlines, you will need a high level of skill and knowledge and well-developed judgment. Not for you the easy decision to put off your flight for a more pleasant day; so you must learn how to cut your decisions finer and still be safe.

Age, Sex, and Intelligence

Age does make a difference. Young people usually learn to fly more quickly, and that's not surprising. First, their recent experience includes listening to a teacher and usually believing what he says. Second, they generally have good sight and hearing. The young person's reflexes are faster, and that usually is an advantage.

But don't be discouraged if you're over thirty, or even over sixty. The oldest student I know of was a seventy-five-year-old woman who learned to fly a helicopter. And many people in their sixties learn to fly and to fly well. It takes longer, but once the mature folk learn, they're just as good as any kid (and often better) when it comes to safe, basic flying, including instrument flying.

Sex can make a difference in a couple of ways. One of these has to do with self-image. There are students, both men and women, who don't put their whole effort into learning. This seems to be because they don't expect to be successful. In the back of their minds a small voice is saying, "You're not going to be able to do this." These people don't work hard at flying, acting as if they don't care, because if they do fail it won't hurt as much if they can say to themselves, "Well, I didn't really try." It seems that more women than men suffer from this self-defeating attitude. But once these halfhearted students become convinced that they really can learn to fly, they progress as fast as anyone.

In another way women seem to have an edge. Women seem to take more naturally to the kind of planning-ahead, anticipation of possibilities, and preparation of alternatives that are so important in so many phases of flying. Women are likely to do a better job of cross-country and instrument flying because that's where those things count the most. And women may get more out of lessons in general because they are likely to prepare more carefully.

Many men feel that planning beyond a bare minimum is not quite manly. They want to spend their time doing, not planning. They enjoy success more if it comes after a series of crises have been met. They like "thinking on their feet," and think it old-womanly to consider all the difficulties that could arise and to plan solutions to problems that may never occur. This approach is often very successful in life, and these men are among those who can afford the expense of learning to fly. But the same attitude which has made them so successful in life may be a problem in their flying.

Do you have to be smart to be a pilot? No. The kind of intelligence that seems to be measured by I.Q. tests won't make a great deal of difference in flying. It's true that the written test isn't easy. In fact, it's pretty tough—so tough that a Princeton physics professor who made a score of 100 on his Private Pilot written test got his picture in the paper. But the test is multiple choice, so if a person can read, write, and do simple arithmetic, he should be able to pass it—if not the first time, then the second, or third, or fourth. As to the flying itself, judgment and a kind of physical awareness are at least as important as knowledge.

I think intelligence and personality are mixed together like scrambled eggs. It is the combination of the two, not one or the other, that determines how we get along in the world.

Judgment depends on the mixture, not on either one alone. And physical awareness definitely doesn't depend on complicated well-greased wheels whirring around efficiently in a really smart brain. In fact, I think that what we call brains may often interfere with accurate perception—with seeing and hearing what is really there.

The identifiable group of students who are most highly selected for intelligence—doctors, dentists, lawyers, and engineers—have a reputation for being especially hard to teach to fly. The difficulty seems to lie in the relationship that's established with the instructor. Many of these people, highly educated and articulate as they usually are, find it hard to accept instruction from someone whom they know to be less well educated and who is not very good at expressing himself. I've found these students are often so busy figuring out everything for themselves that they don't hear much of what I say. They seem to feel that I'm a safety device—a necessary evil that they have to put up with while they teach themselves to fly.

If you are one of this group you can contribute a lot to the instruction process by doing as much extra reading and studying as you can find time for. Remember that your instructor does know all that it's important for you to learn about flying. He has the whole picture, and a well-developed sense of priorities, and he can help you learn if you will allow him to direct your training. Try to trust his judgment.

Don't be afraid to ask questions and to make suggestions. But watch out for a tendency in yourself to pursue unimportant points. Don't let your relationship with your instructor degenerate into a battle of egos. If you find yourself constantly raising technical and largely irrelevant questions that your instructor can't answer, you're probably compensating for an inner unacknowledged dismay you feel because you know he flies better than you do. (I knew a man who used to say, "There are three things I hate: warm beer, wet towels, and people who fly better than I do.") But picking at obscure points, while it may bolster your ego for the moment, will probably have some really unfortunate effects on your instructor. He not only won't enjoy trying to teach you, but he may actually develop a strong unconscious desire to see you frustrated by failure. If you find that you can't really respect and like your instructor—that you are constantly trying to take him down a peg or two—you'll probably both be better off if you find another instructor, one whose superiority in the field of aviation is somehow acceptable to you.

The Way You See Yourself

Self-image is an important part of what we call personality, and variations here make a big difference in a lot of ways. Among my students have been many men and women who always wanted to fly but weren't sure they'd be able to do it. If you question your own ability, let me point out that at least nine-tenths of all accomplishment depends upon persistence, determination, and hard work. Aptitude, talent, even genius, cannot produce achievement without the more pedestrian effort. So it is that your state of mind, your confidence and wholehearted effort, will produce success. It is curious that as we get older we begin to think that if a thing doesn't come easily, we lack the necessary knack for it and shouldn't continue to try. How absurd! Consider learning to walk. Some of us fell hundreds of times. But did we ever say, "Well, I guess I just don't have the knack for this?" Of course not; we just kept trying till we got it. And can you look at an adult and tell by watching him walk whether learning to be a walker came easy or hard? So it is with flying—if you are willing to go on and try again, you will reach a level of practical skill that is as sound as that of the fellow who seems to have been born knowing how to fly.

Some of my best friends used to be among those who were

unsure of themselves as student pilots. These students need a slow and sympathetic start and a lot of encouragement, but as soon as they make up their minds that, after all, they can do it, they begin to put all their energy into learning and they progress as fast as anyone. It is wonderful to see how learning to fly gives people confidence they have never had before, and to see how they begin to bloom. Because this happened to me when I learned to fly, there is a special pleasure in helping it happen for someone else.

Personality Traits and Flying

Your traits and qualities and your characteristic approach to life will be important in your flying. Some of them may go to make you a good student. Others may serve you better as a full-fledged pilot. And there may be others that aren't desirable at all in the air.

There are hundreds of traits that can make up personality. No one person has them all—and different traits suit different circumstances. The careful planner-ahead may not be decisive in an unexpected emergency. The guy who is hail-fellow-well-met and bursting with confidence may make a great living as a salesman. He's an optimist; he really believes that his product is the best, that it will work for every customer if given half a chance, that any flaws or drawbacks are minor if not downright imaginary. His honest belief is catching, and people like to buy from him. But the same general attitude when looking at an uncertain weather picture can be dangerous. There will be a tendency to emphasize the reports that sound hopeful and to discount or ignore those that warn of worse weather to come.

It's easy to see that a reckless attitude is dangerous, but it's also true that a person who tries to think of every possibility and work out a plan for each one can have a problem too.

He may be so involved in minor details that he misses a really important point. Even the obviously desirable and charming trait of being thoughtful, considerate, and aware of the feelings of others can turn up at the wrong time in the pilot-in-command; a lot of accidents have resulted because the pilot didn't want to disappoint a friend.

Maybe you have some ideas of your own about traits you have that will be good, or not so good, for the specific demands of aviation. There are even qualities that help to make a good student that aren't so desirable later when a licensed pilot is on his own. Trust, for example. At first you need to trust your instructor completely, but later you will need to develop a healthy skepticism about everything anyone tells you about flying. You must learn to doubt the Weather Bureau when they tell you about weather, mechanics (not to mention salesmen) when they tell you about airplanes, the Flight Service Station when they tell you about airports, and even controllers when they tell you about traffic.

No one person can have all the best traits for every circumstance, just as no one airplane can be best for every purpose. Aeronautical engineers design airplanes for specific purposes. They engineer out the qualities that don't suit the market the manufacturer is aiming at. The final design represents a lot of compromises among incompatible qualities—short-field capability versus high cruise speed; cockpit comfort versus efficiency in flight; maneuverability versus stability and ease of control. Flaps, and retractable gear, and props with changing pitch are methods of temporarily changing the design of airplanes or power plants to suit the circumstances.

People, too, can temporarily change the balance of their natural traits to suit the circumstances. Qualities such as being always in a hurry, optimism, being aware of what others think, wanting to be thought brave and confident, even overcaution, may be useful to you in your daily life. But hide them away like retracted landing gear when they aren't appropriate in the cockpit. And there are other qualities—skepticism, prudence, deliberateness—that you need to make use of as a pilot. If they aren't appropriate for the way you live on the ground, you can discard them whenever you get out of the cockpit.

The Person Who Shouldn't Learn to Fly

There are a few people whose personalities will never be suited to flying. Most of these people know it and don't even begin to try, but some either don't know it at all or don't accept it. I heard about a man who regularly, about one out of every three approaches, pulled out the mixture control instead of the carburetor heat. Nothing seemed to be able to cure him. The instructor even let the engine stop. When the prop stopped after the landing the student didn't know why it had happened. He was a man who shouldn't fly but wouldn't accept it, and I think his subconscious was causing this silly error over and over in order to tell him to quit. Eventually he was advised to give it up and he did.

Then there is the sad story of the only person I ever advised to quit flying. He was pleasant and eager to learn, and he was not a stupid man. But he was one of those unfortunate people who are subject to hundreds of small misfortunes and accidents. Most of us learn to take a little responsibility for the mistakes of others, looking ahead to see whether someone is about to back out of a driveway, perhaps without looking, or getting out of the way if we see someone coming toward us who is carrying a tray. We've discovered that we benefit ourselves if we plan our own actions to allow for the mistakes of others. But this unhappy man was one of the few people who somehow never learn that. He usually came late to his lessons. Once it was because he'd backed into a car that was parked where no car should have been. Another time he'd had to change his clothes because someone had placed a cup of coffee too close to the edge of the table and he'd spilled it on himself. Things like this happen to all of us once in a while, but they seemed to happen to him almost every day.

In the airplane he never once went all the way through the checklist without a mistake. Now, using a checklist correctly doesn't take brains; a six-year-old who could read would be able to do it, and this man was a lot brighter than that. It seemed that it was something about his characteristic approach to life that was getting in his way. I began to feel that he would never be a safe pilot. We had a long talk about it. He told me his older brother was a military pilot and had said to him scornfully, "You'll never be a pilot." My friend wanted and needed to prove he could do it. So I said, "OK, we'll try one more lesson. Take your time. Do each thing as you come to it. Do the best you can. At least get the checklist right."

That checklist had a fuel pump sequence that went "On" for priming, then "Off" for starting, and then "On" again for takeoff. He turned it on but omitted to turn it off. Before takeoff, when he came to "On" again, he turned it off. The rest of the flight was no better. When we landed we both knew how bad it had been. Still, he wanted to learn to fly "if it killed him." The trouble was, I thought it really might kill him, and maybe others too, so I sadly recommended that he stop taking lessons.

Learning Good Judgment

There is far more to being a good pilot than developing the mechanical skill to fly the airplane well. If that were all there were to it, you wouldn't need the wide-ranging body of knowledge that's tested by the FAA written examination.

You need that knowledge because the Go–No-Go decisions you will make must be based on solid knowledge and understanding of all the factors involved. You can't exercise good judgment without these, and safety depends far more on judgment than on skill. Good judgment can actually make a marginally skillful pilot much safer than a highly skilled pilot whose judgment is poor. That is, a pilot whose decisions always leave a healthy margin between what may be reasonably expected to happen and the situations he knows he can handle is essentially a safe pilot. If he is relatively unskillful, he sets his wind limits, his weather limits, even the kind of field he will fly into and the length of flight he will make at an undemanding level. There are highly skilled pilots who fly regularly in circumstances that call for all the skill they have. I've known some who just didn't leave any margin for the unexpected. The duration of their flying careers was a matter of how long it took the odds to catch up with them.

Traditionally it is said that judgment can't be taught, and it is true that your instructor can't be responsible for everything you ever do in the air. You'll surprise him with your skill and wisdom one day and horrify him with your clumsiness and idiocy the next. But he is not totally helpless in the fight to help you mature as a pilot. He can set you an example that you can feel proud to follow. He can show you that he respects the rules he gives you by following them himself. If he flies by a double standard—doing, himself, the things he tells you not to do, he is demonstrating disrespect for his own rules. The rules I'm talking about are mostly what we call "Good Operating Practices." These are various procedures which add to safety—checking weather carefully or going methodically through the checklists are good examples. These things take time and all too often professional pilots skimp on them. An instructor who is teaching by the "Do as I say, not as I do" method, is quite effectively teaching bad judgment by example. In the instructor-student relationship it is a rare student whose ego is strong enough to ignore the natural desire to be like the instructor.

Fly to Beat the Odds

I've been talking about safety, but let's get a little more specific. We all know that accidents do happen. To some degree we risk injury even when we get out of bed. For that matter, people have been hurt just turning over to shut off the alarm clock. When we get out of bed and start walking around the risks increase. And the risks in transportation are a frequent subject for debate. We are so accustomed to our personal mobility that we don't even think, each time we get into a car, that we are doing something that has a definite statistical possibility of injuring us. Nevertheless, driving a car exposes us to risk, and so does flying an airplane. We try to strike a reasonable balance between safety and utility, and this is where judgment comes into the picture.

The question is, how safe do you want to be? You yourself —whether driving or flying—can reduce the risks to a level far below the ones that appear in the statistics. Accident numbers include a great many that are caused by people who drive or fly when they've been drinking, or when they're angry or upset, or even just plain not up to par. So if you fly only when you're sober and feeling well, the risks you're taking are really much smaller than the statistics indicate, and those are pretty small in the first place. A little care and more good judgment can make the most common accidents impossible for you in your flying. Simply running out of gas has caused a lot of crumpled airplanes. Surely you can avoid that!

If it's so easy, you might ask, why do we go on having these accidents? Part of the answer is plain old fallible human nature. But part of it may be the charter-pilot attitude in so many flight instructors. Most instructors do want to become charter or airline pilots. They admire the men and women who are already doing this kind of work. But the pilot who is in the business of providing transportation often feels a lot of time pressure and economic pressure. These factors, which should have no place in aviation decisions, are nevertheless included, and checklists are done on the run, assumptions are made about gas and oil levels, or weather briefings are omitted. Your flight instructor may emulate his air-taxi pilot friends, you may emulate your instructor, and all of you may put yourselves on the wrong end of the odds.

Finding a Good Instructor

Any certificated flight instructor is perfectly capable of teaching you to fly. There are advantages and disadvantages of learning to fly with an old-timer as well as with a new, inexperienced pilot. The old-timer with thousands of hours has probably seen every mistake you can think of, and he will have developed effective methods for helping you overcome some of your difficulties. On the other hand, I've flown with old-timers who are just plain tired of saying the same thing over and over; they tend to just sit there, ready to back you up if you get into real trouble, and let you teach yourself. The new instructor, though he lacks the instructional techniques gained from experience, may make up for that in enthusiasm and interest. Ideally, you'll find one who has developed the know-how and still retains the interest and enthusiasm.

The most important thing is not technique or enthusiasm but whether or not you can work effectively with this particular person. Of course, no one instructor has all the answers. In fact, there are times when it is helpful to fly with an unfamiliar instructor, no matter how good your own instructor may be. But it does seem to be best to stick to one or at most two instructors. When I was an examiner, giving frequent flight tests, I found that applicants whose log books showed this history generally did better, had fewer hours, and had more confidence. Sticking to one instructor seemed to be more important than the experience level of the instructor.

To work effectively you need to feel confidence in the instructor, and you need to feel that he likes you and wants you to succeed. Fortunately, flight instructors are not like doctors and dentists; you don't have to pay for the time just to meet and talk with them. Instructors have slow days, and when you find one on the ground, not occupied with giving ground instruction, he'll be glad to see you. If there are several airports within reach, it's perfectly practical to go out and spend some time talking about flying and about learning to fly with instructors at each field. Get them to talk about the relative importance of judgment and skill. It's worth taking the time to really look around. You're planning to spend a lot of money and a lot of time, and the things you learn or don't learn may someday make the difference between life and death. Find an instructor whom you can trust and respect and you'll be off to a good start.

2. The Cockpit and Related Training Aids

The most important factor in learning to fly will inevitably be your own traits, abilities, and motivation. Second will be the relationship you establish with your instructor and the influence of his own attitude toward flying. But there are a few other factors that are pretty important too. Some you can't do much about, but it will help to think about them. Let's look at some of the conditions under which learning may take place. Kids in school have green walls, acoustically designed lecture halls, careful insulation from the distractions of the outside world, professionally developed aids of all sorts. You may get pre- and post-flight briefings in favorable conditions like these, but the important learning takes place in the airplane in flight. And what is that like?

Tension and Attention

Most of what is in a cockpit interferes with learning. There is noise, and noise is tiring all by itself. There are so many dials and switches, even in a simple trainer, that you may have the feeling that you'll never get them all straightened out, so why try. Worst of all, for most people being in the air in a light plane brings on a feeling of distress which ranges in intensity from apprehension to terror.

This distress gets in the way of your learning, because your attention is distracted from the job at hand. Imagine that you're at a party, talking to people, relaxed and at ease. Suddenly a madman rushes in with a loaded gun and starts yelling threats. Your attention is concentrated on that loaded gun. You stop seeing, or hearing, or feeling anything else in that room. Being in the air, for many people, is like being in a room with a mad gunman. Their senses are drowned out by distress, and they can hardly hear what the instructor is saying, let alone make sense out of it. Trying to learn to feel a coordinated turn, or to trim, or to judge engine power when you're in this state of mind is a waste of time. Yet these are things you'll be learning.

There is no quick solution to this difficulty. You and your instructor have to wait for the distress level to drop. A patient, calm, and confident instructor can work you through this phase in only a few lessons. But if your instructor shows impatience or scorn, he will only reinforce your distress. If you find yourself still feeling really tense—expecting disaster momentarily—after your third or fourth lesson, talk it over with your instructor. If you don't feel happy about his response to your problem, look for another instructor—one who is willing to accept some responsibility for helping you overcome it.

The instructor's attitude can help you get rid of tension

and apprehension, but there are some other things that are important too. If the airplane doesn't look as if it's well cared for, you'll probably feel uneasy about it. Both the airplane and the engine should be reasonably clean, so you'll be able to spot evidence of oil or fuel leaks. The windows and windshield should be clean so you'll be able to see other traffic. The tires should be good, and there should always be more than just sufficient gas and oil.

Briefing and Self-Briefing

While we may not expect green-walled classrooms, and we do make allowances for the special conditions of flight training, there should be a space set aside in the office for pre- and post-flight briefing. And your instructor should be free to talk to you without interruptions. The ideal briefing area would have a view of the field (most don't—too distracting—but there are a lot of interesting object lessons to be seen out there). There would be books, a compass, a model airplane with movable controls, charts, etc., not to mention desk space—clear, open desk space—and a chalkboard complete with chalk. But even if it's only done in the cockpit with a 9 × 12 lapboard and paper and pencil, the briefing is an important part of every lesson. If your instructor doesn't seem to allow time for briefing in his schedule, talk to him about it.

It isn't enough for him to ask whether you've read your Kershner or whatever manual he wants you to use. He should review at least briefly in his own words the matters the lesson will cover and determine that you really understand the difficult points. And he should tell you precisely what you'll be doing in the air, and what your goals are. If he flies the lesson without any of this, you will miss a lot of what you might otherwise observe. And if he attempts to save you money and himself time by doing it all while you're flying, you will end up doing more flying in order to understand what's happening. An occasional lapse might be OK, but if you never get a good briefing, you can learn to fly better, and more cheaply, somewhere else.

During the briefing, don't just stand there and listen. Questions from you are a really important part of the whole training process. Unless you're absolutely sure that you know what your instructor means—unless it all makes sense to you—it's a good idea to put what he's just said into your own words, repeat it back to him, and ask him if that is what he meant. If what came across to you is not what your instructor meant, he can try another way of explaining. Psychological experiments have shown how important this two-way communication is in instruction. The recipient of instructions can follow them much more accurately if he has the opportunity to ask questions before he attempts to carry them out.

When you're doing the reading, try to figure out answers to your questions by yourself, but don't let questions about the briefing go unasked or unanswered. If you haven't been one to ask many questions up until now, this is the time to begin. When you are willing to admit uncertainty and confusion to the people who seem to have the most knowledge and so ought to be able to help you—in this case your instructor —you will often find that the gap between your understanding and theirs is not so wide as you have thought. Asking questions seldom lowers you in the eyes of those who can answer them. In fact, they will often like you the better for it; for one thing, answering your questions makes them feel both wise and benevolent. And your own self-esteem rises as you become wiser and well-liked.

It's up to you to prepare for each lesson to the best of your ability. It's probably been a while since you've spent any time studying. And this may even be the first time you've *ever* tried to learn something just because you want to. If so, you may not know that all that stuff they told you in school is really true. For example, you do need a quiet place where you can study without interruptions. I've known some people who thought they were "dumb"—that they couldn't learn from books. It turned out that they didn't know that it takes work—that even "smart" people work at it. So don't give up too soon. You should set aside time between flights to review the previous lesson and to think through the one that's coming up. Try to learn the essential points for each lesson beforehand. A good instructor will suggest chapters to read and ways to prepare for each phase of your training.

People who love flying think about it all the time. In our minds' eyes, we see ourselves flying. A lot of us dream that we're flying. I did when I was first learning, and I used to figure that dreaming I was coordinating aileron and rudder and using the elevator properly in the turn was just as good as real practice. As a matter of fact, there's an experiment in educational psychology that bears this out. They took a group of basketball players, evenly matched in skill, and set half of them to practicing shooting baskets for an hour. The other half only imagined they were shooting baskets. After the hour the two groups were tested, and the ones who had been imagining had improved *more* than the ones who had really been practicing. If you love flying, you're doing some of this imagining already. But even if you're learning only because it will be a useful skill, you might want to try it. It beats paying for an airplane, and it will make the time you do pay for more valuable to you.

The Problem of Negative Transfer

Imagining yourself doing the right thing can also help with a problem that troubles everyone. This is the psychological phenomenon called transfer. Almost all of us have experienced this at some time. A touch typist using an unfamiliar and slightly different machine will hit the wrong key for the back space or the margin release. A new car with different door handles, or locks, or even radio switches will slow you down. Your hand goes to the familiar spot and then, when the habitual response, the learned reflex, doesn't work, the brain steps in and the job gets done. Until your hand has learned the new habit, you'll be a little slower than someone whose hand has no habit to unlearn.

Unfortunately, there are several ways in which the design of modern airplanes encourages the transfer of wrong responses, and so slows down your learning of the right ones. If you are accustomed to a throttle on a boat, the one in the airplane will seem to you to work backwards. In the boat your hand has learned to pull out for more power. Now you have to learn to push in. If you remember to associate the throttle with an accelerator in a car, it may help you to remember which way it goes—in for more power.

And for most people the rudder pedals seem to work backwards. Steering with our feet isn't something most of us have done much of, but even so, pushing forward with the right foot in order to turn right doesn't seem to come naturally to us. Skiers, in particular, have learned to use their feet in the opposite way and may have extra trouble learning the new response.

Transfer can even interfere with learning to read the instruments correctly. I had one student who was making a connection in his mind between the airspeed indicator and a dial he used constantly in his work. On the dial at work, motion in a counterclockwise direction meant a higher reading, so in the air he responded by raising the nose, which worsened the situation. When we figured out the reason for his error, his discouragement and his embarrassment at what seemed to him to be rank stupidity vanished and he was soon able to learn the correct response.

Of all these wrong-way transfers the one that gives people the most trouble is the wheel. When most airplanes had a stick which banked the airplane there was no difficulty, but today the wheel or "yoke" is the usual form of the control. The problem seems to be that almost all of us drive and have learned very, very well a hand response to unwanted turns of even fractions of a degree. But this is the wrong response in the airplane. The right thing to do is to use a little rudder pressure.

The movement of the wheel, if it is very slight, isn't enough to cause serious trouble, and it can happen hundreds of times over the first few hours of flight training without either the instructor or the student even noticing it. All this unnoticed and unconscious practice reinforces this wrong response. Then one day, cruising in rough air, or working on takeoffs and landings in a gusty crosswind, you may find you have a really troublesome habit getting in your way. The quickest way to overcome wrong reflex responses is to slow down your responses until your brain has time to get in there before it happens and prevent that wrong movement from occurring at all. In this way you will unlearn the wrong and learn the right.

Airports and Traffic as Learning Factors

The airport where you fly is another factor that can make a lot of difference when it comes to both the time and the money you'll spend learning to fly. If you have a choice, here are some things to think about. My personal choice for an ideal field for training would be a long, wide, smooth, flat *grass* field with no traffic. The next best has long, wide, hard-surfaced runways and no traffic. There are a few old WW II training fields which fit this description, but most are in the middle of nowhere and a lot of them are abandoned. Some county-maintained fields fit these specifications. If there's one near you, you're lucky.

You're more likely to be flying out of 1) a controlled field with wide, long runways (and a lot of traffic), or 2) a field of reasonable length with one narrow, hard-surfaced strip (and little traffic), or 3) a small grass field, probably with obstacles which in effect shorten the landing area still more (and no traffic). Now this will sound like nostalgia casting a golden glow over the dark past, but it really was a lot easier to reach the point of that first solo when we were flying in Cubs on wide grass fields in the good old days, than it is now in the new easy-to-drive nosewheel beauties on the expensive paved runways. Part of the difference had to do with the advantages of a big grass field. Directional control was easier on the ground, and of course there was no center line, so aiming at the field didn't have to be so precise. Furthermore, a cross-wind could be changed into a head wind simply by angling the takeoffs and landings across the field. A hard-surfaced strip just doesn't allow for such flexibility.

Traffic is another problem. If the only time you can fly is weekends and evenings when everyone else is free to fly, you are probably going to spend a lot of time in the airplane sitting on the ground waiting for other traffic—especially when you start working on takeoffs and landings. The best solution to that problem is to schedule your flying for off hours. There are several possibilities. Weekdays are best and have this extra advantage over weekend and evening flying: Since there is less competition for the time, it will be easier to reschedule for the same time the next day, if you hit a day of bad weather. A lot of fields have a traffic lull between five and six on summer evenings. Early mornings are also good, and even if it means getting up very early on Saturday and Sunday (which in turn means going to bed early on Friday and Saturday), it will be worth it to avoid heavy traffic. If none of these are possible for you perhaps you can make it before work during the week. If you are stuck with Saturday and Sunday, my experience has shown that even late morning is better than afternoon, and that Saturday is quieter than Sunday.

The Weather as a Learning Factor

When you're taking flying lessons, you soon find that you are thinking very differently about weather. The things you hoped for as a groundling—sun and pleasant temperatures—aren't important to a pilot. What you're looking for as a pilot, and especially as a student, are a clear horizon and smooth air. There are parts of the country, including the east where I've done most of my flying, where a clear horizon is rare. Really smooth air is hard to find too. A sunny day, perfect for a picnic, hot but with a nice breeze, is not a good day for instruction. The air will be rough. The airplane will be bouncing around, and will seem to be getting itself into attitudes that have nothing to do with the way you're using the controls. That can be confusing.

You'll learn more and learn it better if you have a clear horizon for at least the first few lessons. If you are flying in haze with no horizon to give you a level reference, you will almost certainly begin to use the artificial horizon (which the FAA prefers to call an attitude indicator), and you may stop even trying to use the horizon outside. And that's bad, bad, bad! If you're going to drop either one as a reference, drop the artificial horizon. It's dangerous if it's overemphasized in VFR (Visual Flight Rules) flying; here's why:

When you keep your eyes outside, where the real horizon is, you have a good chance of seeing both traffic and weather before either one can hit you when you're not looking. But once you've learned to use the instrument, it will seem much easier to be sure that the wings are level, or to be sure that you have the bank you want, if you refer to the instrument in the panel. And if you start that way, relying on the instrument, you may never develop ease and confidence in using the natural horizon. So if you do use the artificial horizon because of poor visibility or some other reason, make a conscious effort to use the outside horizon every time you have reasonably good outside references.

Wind direction won't be much of a problem for the first few lessons when your instructor is making the landings. But if the wind is causing rough air, that's good reason to cancel. Rough air will mask your errors, or exaggerate them. Neither you nor your instructor will really know exactly what's going on between you and the airplane. The whole point is that you are trying to learn correct reflexive habits. That is, you want to learn to move your hands and your feet to control the attitude of the airplane without having to think about it. In smooth air you'll see that the left aileron will put the left wing down and the left rudder will start the turn smoothly. The nose attitude will be affected by the bank and you'll need a little elevator. When you do the right things, you will see the expected effects. But if the air is rough, it is possible that, because of a gust or "bump," none of the normal things will happen. That means you won't be rewarded for correct behavior, and that means it will take you longer to learn.

Wind direction as well as wind speed become important when you start doing takeoffs and landings. You can't change the weather, but selecting the light traffic hours, early morning or evening, will usually give you the best wind conditions too. A lesson in calm smooth air at 7:00 A.M. can really give your day a beautiful start, though you may find your instructor doesn't show the same enthusiasm you do. If you're thinking, "But, I'm going to have to learn to handle rough air, and wind, and poor visibility sometime," I'll agree you're right. But learn the basic control reflexes first. You'll feel more confident in the early stages; you'll make better progress; and when you do start working with rough air, wind, and poor visibility, you'll learn to handle them more quickly and easily than most people do.

Some Effects of Financial Pressure

Financial pressure can be a very important factor in your progress toward a pilot certificate. If the instructor and the people he's working for have financial problems, you may be encouraged and expected to keep appointments and fly lessons when conditions are far from ideal. The instructor won't allow rough air, haze, or wind to deter him. Minor problems with the airplane—dirty windows, low tires, a bad fuel gauge—will probably be ignored. Taking the time out of the schedule to fix them would mean lost revenue. But in the meantime you are getting a powerfully bad example from the very person who is your model as a pilot. He's teaching you that routine safety procedures can be ignored if time or money (or, presumably, other personal pressures) make it expedient to do so.

Financial pressures may also lead to lessons being given when the instructor or student isn't feeling up to par. And another insidious factor may contribute an effect here. There is some of the "iron men in wooden ships" sea tradition in aviation. A young instructor, a would-be airline, air-taxi, or company pilot, has heroes who probably aren't men who will die in bed. They take pride in flying whenever scheduled, rising above such ordinary human woes as hangovers, toothaches, or head colds. Of course, the fact is that there are few indeed who can do a really good job, even at what they know well, when their bodies are in bad shape, and I doubt that any of these few are neophyte flight instructors. So between money needs and the iron man syndrome, a lot of questionable training flights are made.

The first time I experienced the out-of-it instructor was on my first (and only) dual cross-country. It was hazy; the instructor clearly didn't really want to go, but he told me to go up around the pattern and take a look. If I still wanted to go, he'd go. I wouldn't have another chance for at least a week, so I looked and told him I'd like to go provided he'd let me take a road map as well as the aeronautical chart. He agreed and we set off in the Cub, I in the front, he in the back. He gave me some help in the first five miles or so, and then I settled down to the course. At the first "point of intended landing" I turned around to ask him something about entering the pattern and found he was sound asleep. Well, it *was* early Sunday morning, and besides I was flattered at what I thought was trust in my ability to navigate. Now I know that I could have flown thirty or forty miles off course and he would have discovered his position within a minute or two; and I also know how important that extra pair of eyes might have been.

Recently at a well-run, modern flight school I heard a young instructor saying that he hoped his night flying student wouldn't show up because he was so tired. The student came, saying he had a bad cold. The instructor ignored that, though he double-checked the wind, clearly hoping it would be strong enough to justify canceling. It wasn't, so off they went. That student with his bad cold couldn't have learned from his tired instructor more than half of what he would have if they'd both been in good condition. The lesson, in effect, cost him double price. More important in the long run, he learned, as the instructor himself had evidently learned before him, that physical condition is not a Go–No-Go factor in aviation. That was an expensive and dangerous lesson. That kind of demonstration contributes a large part to the numbers of accidents that involve pilots who are tired or ill when they begin a flight.

Financial pressure on you, the student, will have other effects. One may be that you'll have to quit periodically while you balance your accounts. This can be discouraging, but it's not a hopeless situation, and you'll find that you don't spend more than an hour or two at most getting your proficiency back. I've had students who were struggling with takeoffs and landings and making little progress come back after a six-week layoff and do better than before. Perhaps they were practicing in their heads, but I've always thought it was because, not expecting much of themselves, they were more relaxed, though still alert.

There is another effect of financial pressure on you—an effect of the high cost of learning to fly—that most disturbs me when I'm thinking about the training of safe pilots. What happens is that a kind of unspoken conspiracy is formed between you and the instructor, and the owner of the flight school, to "get you through" in the least possible time. It is insidious, and it is general throughout the industry.

The Minimum Time Trap

Before taking the flight test, the applicant is required to have a minimum of forty hours of time in the air logged in his record, including twenty hours of solo practice. That has been the minimum requirement since way back before there were instruments and radios for the pilot to deal with. It is now a ridiculously low minimum—something like saying you have to have thirty minutes behind the wheel of a car before taking a driving test. But since the figure is set forth in the books, a general, industry-wide, minimum standard is set as the goal for the proficiency of the new private pilot. There are no objective gradings of proficiency, so the standard of excellence during training is often established on the basis of how quickly the student can meet the minimum standard of proficiency—how much, or little, time he needs to pass the test.

Let's see how that works out. A very apt student, one who, in addition to unusual natural aptitude, also has some extra experience in the air flying with a friend or relative, might be able to meet the minimum requirements on the private flight test when he has logged forty hours. If he is instructed specifically for the test, and if he takes the test with an inspector whose method of testing and special concerns are known to the instructor (both circumstances are frequently the case), this will be quite possible.

Here we have a potentially outstanding pilot, who becomes licensed when he is barely proficient. Despite the fact that it is always said that the private certificate is a "license to learn," very few pilots continue to improve—certainly not as rapidly as when under supervision. Many, according to one FAA inspector, never fly as well again. On the other hand, if this apt student were to work toward certification for the same length of time (fifty-five to seventy-five hours) that it takes the average or slow student, he will begin his unsupervised career as an outstandingly skillful pilot.

What we have is a situation where everyone graduates with a D-minus—some because they have a lot of trouble with the material to be mastered, some because they don't work long enough to do as well as they might. So, if you want to be a really sharp private pilot, plan right now to fly a minimum of seventy-five hours before taking the test. Give yourself the time you need to polish and perfect your performance. With hard work and a cooperative instructor, you'll have a fair chance of becoming really good.

Flying Is Well Worth the Cost

Learning to fly is expensive and getting more so all the time. When I started flying, the Cub rental was $8.00 per hour and the instructor cost me $3.00 (awake or asleep). Today it may cost you more to get an instructor when you already own your own airplane than it used to cost to rent the airplane complete with instructor. And still most instructors don't earn as much as a cocktail waitress or even a top-notch cleaning woman. If instructors were paid enough to hold the pilots who are really good teachers, it would be even more expensive to learn to fly; but we would have a better aviation world.

Once you get past the instruction phase, flying comes off pretty well compared to other ways in which people spend money. It costs no more to own and fly an airplane than to own a nice boat. An airplane is less trouble to care for than a boat. And flying unlocks doors that will be forever closed to you without it. An airplane extends the horizons of your world. You can take day trips and weekend trips to places that would otherwise lie impossibly far in time or space, or at the end of hazard-filled and unpleasant hours of highway driving. In flight, your mind, life, and spirit will be stretched and for that the price is cheap indeed!

3. The First Lesson, Part One—On the Ground

The first lesson is especially important. You may remember it only as a kind of blur of impressions, but it is likely to influence the way you think and feel about flying for a long time to come. There are a lot of things about that first lesson that you can't control—the weather, the airplane. But you do have control over some things that are very important—your own state of mind and your own attitude toward the job of learning to fly. Right now is the time to begin accepting responsibility for the way your training goes. You can start by getting some basic reference materials.

Basic References

The book I start every student with from the first lesson, or even before, if I can, is William Kershner's *Student Pilot's Flight Manual.* This is sound, thorough, and readable. It has hundreds of helpful illustrations and has an index, an essential in a book to be used for reference and one which is too often lacking in the otherwise excellent government publications. You can save yourself time in the air as well as money by getting and studying this book. It comes from Iowa State University Press in Ames, Iowa, and most airports carry it. If your instructor doesn't ordinarily use this book, you can select and read relevant chapters on your own as you go along. The Piper or Cessna courses do not substitute for Kershner, nor do the government flight training manuals.

Another basic reference I like my students to have is *The Private Pilot's Handbook of Aeronautical Knowledge.* This is a government publication covering all the basic information except regulations. From it you will discover what the government thinks is important to know. Of course, this is what the government test will cover, so a good home study course can be based on this book.

You will also need the Federal Aviation Regulations which you will be required to know, and these are available from a number of commercial sources. Most airports carry one or more of the books or booklets containing them. Of course, they are available from the government too, but that can be complicated, costly, and time-consuming.

Don't fail to get a complete set of Exam-O-Grams, another government product. They are produced in response to obvious need on the part of students and pilots. Some specific misconceptions and misinformation are found to cause trouble on written tests, oral examinations, flight tests, or to be a factor in accidents. Exam-O-Grams are designed to explain information found to be especially difficult or confusing. They appear irregularly, are one to four pages in length, and have been coming out since 1960, so there are a lot of them. The government will send you those that are still current (some dealt with now-superseded regulations, or radio frequencies no longer assigned, or other changing information) at no charge, and they'll put you on the list to receive new ones as they appear. All together, these sheets make up a well-written and illustrated textbook, straight from the horse's mouth, covering much of the most crucial material. Some have been included in the *Private Pilot's Handbook* mentioned above, but send for them anyway so you'll be on the list for new ones as they're issued.

Finally, although navigating won't be your responsibility right away, it would be a good idea to get an aeronautical chart of your area now.

A Bird's Eye View of the Pilot's Job

The immensely complex and demanding job of flying can be boiled down to a list of skills and abilities you'll have to develop. There is a natural development and order to learning to be a pilot. The whole list is here so you can see how each training phase fits into the whole picture. In your first lesson you will tackle only parts of a few of these elements. Training and practice will take you through the list at a rate that is geared to your individual needs and circumstances.

Here are the things the pilot must be able to do:

1. Recognize the basic aircraft attitudes for the "four fundamentals": straight and level flight (cruising), climbs, glides, and turns.

2. Achieve and maintain the desired aircraft attitudes through the proper use of controls.

3. Coordinate power and attitude and configuration (flaps, spoilers, etc.) to control altitude and speed.

4. Use the "performance instruments"—airspeed indicator, altimeter, and compass—to check on the correctness of coordination of power and flight controls.

5. Fly a desired ground track, with or without wind, while maintaining assigned altitude and/or airspeed.

6. Judge and control path and altitude and speed relative to the ground so as to effect takeoffs and landings.

7. Perform these tasks while constantly searching for other air or ground traffic, and checking for indications of possible emergencies due to weather or other difficulties.

8. Control airspeed, altitude, and heading by reference to instruments alone.

9. Navigate by reference to landmarks, charts, compass, and radio aids.

10. Use the radio for efficient communication in the present information (FSS—Flight Service Station) and control (ATC—Air Traffic Control) systems.

11. Find the information necessary to the safe conduct of a cross-country flight.

12. Understand air density (affected by temperature, pressure, humidity, and elevation) and wind velocity and their effect on aircraft performance with particular reference to takeoff, climb, glide, and landing.

13. Understand general weather elements and their effects on the probabilities of successful completion of a given flight.

14. Find and properly apply information relating to the performance of your airplane, including fuel system, lubrication, electrical system, radios, hydraulic system, loading, and performance data.

15. Anticipate, recognize, and assess unusual, hazardous, or emergency situations, and take the proper corrective action.

A standard sequence is usually followed in presenting the maneuvers, information, and procedures that will bring you to the private license. This list does not parallel that sequence precisely, but you will find that some of the maneuvers you'll be asked to practice will make better sense if you think in terms of the whole list. Your first lesson, for example, covers some (not all) elements of 1, 2, and 3 in the list and should also include looking for other traffic which comes under 7. The responsibility for the rest of the job lies entirely with your instructor in this stage—such things as navigating back to the field, constantly monitoring for potential hazards, and using the radio.

Sometimes there are advantages in presenting elements in nonstandard order. A lot of people learn to navigate before they handle the controls even once, especially wives, husbands, children, parents, and friends of licensed pilots. There are people who, for one reason or another, do all the book work first, without any flying at all until after they pass the written test. And sometimes people learn to fly by instrument references in ground trainers before they set foot in a real airplane. However, the rest of this book generally follows the course of normal training, ultimately covering all the list, as well as proposing ways you can practice, explaining underlying principles in nontechnical terms, and discussing priorities in terms of flying safety.

Pre-flight and the Controls

There is a trend toward doing a lot of in-office briefing, but the very first briefing can be done nicely out by the airplane while the instructor works through a pre-flight. This pre-flight inspection of the airplane before every flight is one of the ways that pilots can make flying safer than driving. Some instructors skip this demonstration on the first lesson, feeling that the important thing is to get you into the air. But you'll be missing the opportunity to learn a lot if you don't walk around the airplane reviewing what you've studied in the manual.

If you watch the aileron move while the instructor moves it with the control, you'll be more aware of what is happening when you fly. Notice how small is the deflection of the control wheel in the cockpit necessary to produce deflection of the control surfaces on the wings. Look at the angle of maximum deflection upward, and notice that this angle is greater than the corresponding downward deflection. This difference is set intentionally to help to compensate for adverse yaw. Like the design characteristics which help to compensate for power effects, this works best at normal cruising speeds.

"Deflection," "adverse yaw"—I can almost hear you thinking, "Why can't people use ordinary words?" Sorry, but you really need to get comfortable with aviation's special vocabulary. The ideal way would be to have someone read the rest of this book aloud to you while you're seeing, feeling, and doing what's being described. But that may be pretty hard to arrange, so I expect you'll read some, go and do what you've read about, maybe reread that, and then go on. You won't remember all of what's suggested. Why not make a few quick notes on a sheet you carry into the airplane with you?

Getting back to the pre-flight and deflection, look at the flaps and see how they, in effect, change the design of the wing temporarily. (See Kershner, page 92.) Later, in the air, you'll be able to feel how flaps slow you down. And that demonstration may help to make it clear how aileron use causes adverse yaw.

When the instructor gets back to the tail section (or "empennage"), he can give you a clear explanation of how the horizontal surfaces there work—the stabilizer, elevator, and trim. You'll begin to understand changes in elevator pressures and required trim when you understand that the horizontal surfaces back there don't help to hold the airplane up, but instead exert force in the same direction as the weight of the airplane. It's true that a lifting tail surface would be more efficient, but such an airplane is so unstable as to be virtually uncontrollable. (See figure 9–9, page 54, in Kershner.)

Gas and Oil and a Couple of Cautionary Tales

There are some suggestions about the pre-flight procedures that may be helpful. Make the check of the gas and oil first; if you find either one is low you can taxi to the pump or call the truck and go through the rest of the pre-flight while the airplane is being serviced. Be very meticulous about draining gas and checking it for contamination as well as for the correct color (indicating octane). It is usual today to drain gas into a small plastic container, but this method has disadvantages. If only very little dirt or water is present it will be hard to see and may stick to the bottom of the container. When you then drain some more gas and check it a second time, it will be almost impossible to be sure whether or not you have gotten rid of all the contaminant. The best way to see the gas and all it contains is to drain it into your hand. Even one or two tiny drops of water or bits of dirt will stick to your skin, the water standing up in beads. You can wipe off any contaminant you find, drain again the same way, and have confidence in what you see. I know the gas is unpleasant on your skin, and in the winter it will feel painfully cold, but it is the best, and I believe the only, way to be sure about your gas.

I've had two very interesting and educational experiences

with water in gas. On one occasion a gas cap was left off the tank of my Cessna 172 overnight during heavy rain. Not surprisingly I found water in the tank when I drained it before flying the next morning. I drained the tank until several ounces were pure gas. Back on the ground after the flight, I drained both tanks and the carburetor sediment bowl again. This time I got about a teaspoonful of water from the one affected tank. I began to drain the tanks and the sump both before and after every flight. (Usually I drain only before a flight.) I found a few drops of water in the affected tank every time, but never had any in the carburetor sediment bowl. Finally, after three weeks of draining everything several times a day, I stopped finding water, and reverted to the practice of draining only before a flight.

The second time the Cessna got water in the gas, it could have resulted in a serious accident. A student had the airplane and got gas at a nearby controlled field while on a solo flight. Unbeknownst to him, the gas truck had not been used for at least several weeks. My student paid for the gas and started the plane. After getting his clearance he began taxying out for takeoff. While he was still on the taxiway the engine quit. He finally got pushed back to the ramp and investigation showed there was rusty water in the gas. When I flew over with a colleague we found the plane down at the end of the ramp on the grass, my student and a line boy draining the tanks through the quick drains. When several quarts had finally run clear (and that took a while) and the carburetor sediment bowl was running clear too, I got in and flew the airplane home with no further indication of trouble.

The next morning I found a few drops of rusty water, and then the gas was clear. But during the first takeoff run that morning the engine faltered. I was with a student who was already very nervous about flying, and though we aborted the takeoff while still on the ground—there was no real problem, and the engine never quit altogether—it was enough to end his flying career for good. After the aborted takeoff I decided that I wouldn't feel sure of the airplane again until the tanks were flushed and the carburetor was cleaned. That was the only time in six years with that 172 that I ever lost revenue for a whole day for unscheduled maintenance. I was not able to get any compensation from the operator who had pumped the rusty gas, though they did finally refund me the cost of the contaminated gas. The moral of that tale is to be careful about gas from trucks, even at big, busy, and normally well-run operations. And maybe, too, it would be well to drain tanks after getting gas, though there's some doubt as to whether the water will show up immediately. If you have reason to think your gas is contaminated, it's safest, albeit expensive, to flush the tanks and clean the carburetor.

It is also important to check the level in the tanks visually, and check what you see against what the gauges say. You not only want to be sure that you have enough gas for the planned flight and more, you also want to know how it's distributed in the tanks so you can plan your fuel management in flight. If you're flying a Cherokee or any other plane in which the gas load can become badly unbalanced, be very careful not to let that happen when you're flying, especially in the early lessons. Start out on the fuller tank, and if you start with both tanks full, switch tanks after the first half hour, and every hour after that. In the Cherokee the heavier wing begins to be a noticeable problem after about a half hour. Soon you'll find you have to hold constant pressure on the aileron control just to fly straight and level, even in smooth air. When your flying is more difficult, it is also less safe than it should be. You grow tired sooner. You can't be physically relaxed, as you need to be in order to feel what's happening. So head off these problems by keeping the gas load reasonably even.

More About Pre-Flighting

I've seen some surprising things overlooked on pre-flight inspections—bald tires with cord showing, a prop which had the tips curled, a missing gas cap. People have been known to miss more astonishing things than those. Every once in a while someone attempts to take off with no rudder at all. Hard to believe, but true! Of course, you know you need a rudder, and you know where it is, but I've found a lot of applicants on flight tests who don't know what the static port is, or where to find it. Without a clear static port your altimeter won't work, and your airspeed indicator and vertical speed indicator will be affected too, so be sure you know where it is.

I came to the pre-flight rather late in my career. At the field where I learned, the two instructors who were partners in the business used to get the airplanes out of the hangar in the morning, check them carefully, and fly each one around the field once. They did not teach pre-flight inspection procedures—what used to be called the "walk-around"—until just before a student was to go for his flight test, apparently feeling that ten or fifteen pre-flight inspections a day were too much of a good thing.

Anyway, one day I set off on my first solo cross-country, a route of my own choosing, with landings at three small fields. It was exciting to be on my own, and the first leg went well. On the approach to the second field, however, I spotted some wires barely in time to avoid them—the poles were hidden in an orchard. At least the landing was OK, but when I got safely on the ground I found out how shook-up I really was. The young instructor there told me they'd had three accidents due to those wires in the previous year alone. I had a Coke, wrote shakily in my log book, and chatted with the instructor who was especially interested to see a woman flying. It was a little less common in 1958. He had tried to teach his wife, but she got sick every time she got into the air. I remarked somewhere along the way that I had never done a walk-around, and he offered to show me how it should be done. We were both surprised when he found a flaw in the belly of the Cub. There was a hole in the fabric which had been covered with tape, and now the tape was coming loose and hanging down. He felt this was definitely an unairworthy condition, and he got the mechanic to put on a proper patch free of charge.

All this made me pretty late arriving at the next field, Sky Manor, which had a reputation for being hard to find. My husband with our three children had flown up there to surprise me, and of course I was so late that they had thought I was lost. When I finally got home (I had called to let the field know I'd be late getting back), my instructor just said, "Well, we'll have to make sure to send you out with a plane whenever we have one that needs fixing." He was kidding, but also he was teaching me not to be concerned about careful maintenance. His casual attitude did not rub off on me, however; I do think the pre-flight is important and I never skip it, even if I'm *sure* nothing could be wrong.

In the Cockpit for the First Time

When you climb into the cockpit after the pre-flight, the first thing to do is to make sure you feel comfortable in the seat. To make good use of your instruction in the air you need to be concentrating on the effects of the controls and on the look, the sound, and the feel of the airplane. To do this you need to be able to sit relaxed with the weight of your legs resting on your heels on the floor, and your toes resting lightly on the bottoms of the rudder pedals. If you need cushions, either to reach the controls comfortably or to see out, now is the time to get them. Some old-timers scorn the desire to see over the nose. In the wonderful old planes they used to fly they couldn't see out in front of them. Well, OK, but they had the compensations of the open cockpit, and the feeling that flying was an adventure. Every hardship somehow proved their worth. It's been said that adventures arise out of poor planning. Pilots today prove their worth by avoiding adventures.

You'll be more aware of the sights and sounds and other sensations that are so important at this stage, if you don't try to make use of all the instruments just yet. This very first lesson is the time to start learning to judge your airspeed and your power setting by ear. If you're concentrating on the needles on the dials, you won't be paying enough attention to what you hear. But you probably will want to use the altimeter, so find it on the panel and be sure you understand how to read it.

Before the instructor starts the engine he'll let you experiment with moving the controls. The pressures you feel now aren't the same as those you'll feel in flight, but you can take the first step in getting the feel of flying by moving them now, without the flight pressures of air flow over them. Then later you can compare what you feel when the plane is flying. The wheel should be held lightly, the way you would hold a live bird—tight enough so it can't get away, but not so hard you'll crush the life out of it. This is important. For one thing, holding the wheel in a death-grip will be exhausting. And also, if you habitually hold the wheel very tight, you won't be able to feel small changes in pressure against your hand in flight. These changes can tell you things you need to know.

Checklists

Your instructor will probably go through the various checklists—pre- and post-start, pre-takeoff, and pre-landing—in the first three or four lessons. Listen carefully, and watch, so that you'll have an idea of how it's done when he's ready to have you do it. For many years I had my students begin to use the checklist on the second lesson. I used to congratulate myself on my patience. But after trying both methods I now believe that it is better to wait. There is a lot of expensive time when a very new student tries to go through the checklist. And I think, too, that this responsibility in the first few lessons simply introduces a source of tension into the training session, and adds unnecessarily to the general confusion. Besides, you'll have plenty of opportunities to practice these procedures even if you don't take on the responsibility until the fourth or fifth lesson.

The pre-start portion of the checklist is short and it is true that if you miss something here you can pick it up later when you do the pre-takeoff check. Some people treat the pre-start list very lightly. I once saw a professional and highly regarded air-taxi pilot spend almost half an hour trying to get a recalcitrant engine to run. He was so disgusted when he finally realized he'd left the mixture control in the idle cut-off position all along, that he got out of the airplane and strode off muttering, and didn't reappear for several hours. There is a minor controversy about one item that appears on most pre-start lists—shout "clear." Some people feel that in a modern plane, one in which the pilot can easily see over the nose, this precaution, which used to be necessary for the sake of bystanders, is absurd. I cannot agree. Even a quick and careful look around before hitting the starter does not guarantee that there is no one directly in front of you under the prop. Unlikely, I admit, but not impossible. It could be a bystander who noticed that you had left a chock under the nosewheel. Anyway, do shout "clear," *and* wait long enough so there's time for an answer if it isn't clear.

The engine run-up and the rest of the pre-takeoff procedures are at least as important as the pre-flight inspection in contributing to the safety of your flight. For one thing, an inadvertent omission earlier may be picked up here. The written list should be covered slowly, carefully, step by methodical step. And since you're going to use it hundreds or thousands of times in your flying career, make sure the list is a good one. A lot of checklists aren't organized in a logical sequence, and a lot of them are too abbreviated. Of course, you ought to know what "Check controls" means, but I've seen too many people leave out the check of the rudder if their list isn't specific. And the blanket phrase, "Flight instruments," is an invitation to omissions too.

The same old macho, iron-men-in-wooden-ships thing rears its ugly head when it comes to checklist use. Some old-timers and their young imitators feel that somehow the use of a good thorough checklist is an admission of weakness. These men feel that they are the elite, and they look down on those who use careful checklists. It all comes down to how safe you want to be. For them, making flying safer—more sure—makes it less meaningful. Everyone approaches flying in his own particular way, but the costs are high for this way of being a pilot.

First, the pilots themselves pay with their lives when their humanity catches up with them. What might have been only a minor error, such as anyone will make once in a while, sometimes turns out to be fatal because they haven't backed themselves up with sound procedures and good operating practice. If their own lives were the only cost, it wouldn't bother me so much. Not that I'm heartless, but I do feel that people have a right to do as they please with their own lives.

Unfortunately others pay too—their families, friends, and passengers—in mental or physical anguish if not with their lives—and that does bother me a lot. And, more selfishly, I am angered because all of us who fly pay too, little by little, year by year, in higher insurance rates and more restrictive regulations.

I do think that any licensed pilot should be able to carry out a thorough safe check without a written list, if that becomes necessary, but I also believe the list should be used as a regular thing. If a new model or a new piece of equipment requires that you delete or add an item, or change the order of the items, don't make such changes mentally. Write out the whole new list exactly as you want it. And never skip around in the written list; that too opens a clear path to error or omission. Doing the checklist single-handed, as you normally will, may mean that when you look down at the list after checking one item, you'll inadvertently skip the next item. You can avoid that by keeping your finger on the list and by rereading the item just covered before going on to the next.

It is just those times when you're most tempted to skip the written list—when you're in a hurry—that it's most important to use it properly. That's because if you're in a hurry it probably means that something else is on your mind. When something else is on your mind, you're likely to leave out something important if you go through the checklist by memory. And it is during the pre-takeoff procedures that you are most likely to feel pressure to rush through the checklist. There may be another airplane waiting behind you, or a lull in heavy incoming traffic. You'll need to discipline yourself to resist the temptation to hurry. If you should be unlucky enough to leave out the one thing that happens to be in a hazardous condition, your haste could be fatal. Consider the following true horror story.

I knew a man who owned a 172 and had flown it regularly for ten years or more. Then one day he had a fatal crash during takeoff. It turned out that the control lock inside the cockpit hadn't been removed. Now, if some instructor had said to him, "You ought to use a checklist," he would probably have said, "You're crazy. I could do that check in my sleep." But every checklist includes an item that covers free and correct movement of the controls, and he'd be alive now if he'd used a good checklist. No matter how many hours you accumulate in the air, no amount of familiarity and experience will ever justify your skipping the checklist.

Taxying

After the pre-start list is completed, the engine is started, and the post-start list is taken care of, rest your hands and feet on the controls while the instructor taxies. Notice that the rudder pedals are never completely still during taxying, any more than the wheel of a car is completely still during driving. A lot of students seem to have trouble taxying simply because they're unconsciously trying to find the one position of the rudder pedals that will keep the airplane rolling straight. Of course, there isn't any such position.

Here's a useful trick that a lot of people don't know. You can use the shadow of the wings to know where your wings are in relation to obstacles around you on the ground. The most common obstacles you'll have to worry about are the wings of other airplanes. Watch those shadows; if the shadows of the wings don't overlap, the wings can't possibly touch each other, let alone crease or crunch each other. If the shadows do overlap, it is still possible that the wings will remain apart if they happen to be at different heights. But usually we're concerned about other wings at approximately the same height. Keep the shadows apart and you won't have to worry.

Also, while the instructor is taxying, look around and try to fix in your mind's eye the look of the horizon when the plane is sitting level. Look out ahead of the plane and at the wing tips. Flying isn't like driving a car, where most of your attention has to be focused on the road. You need to begin to learn to look in unfamiliar directions—up, down, and out to both sides. Move the control wheel; the controls will feel different now that the airplane is moving, though the pressures will not yet be quite like those in flight.

Safety Starts on the Ground

All of this first-lesson, before-you-get-in-the-air stuff bears on essentials of flight safety. Mostly it comes under items 14 and 15 of the list of pilot tasks—knowing the airplane, and watching for trouble. The chronological place for this discussion is right here where it is, but all of this is still the instructor's responsibility at this stage of your training. If he's like me, he'll expect you to be able to do a thorough pre-flight, one he can depend on, by about the fourth lesson, and you'll begin handling all the checklists by then too. Some people use a checklist for the pre-flight inspection and that's not a bad idea either.

You may wonder why I haven't said anything about using the radio. If you're flying out of a controlled field or are reporting takeoffs, departures, approaches, and landings to unicom (radio facility on uncontrolled fields), this should also be the instructor's business for now, and I'll give you my views on radio work later. Right now I don't want you distracted from getting to know and like the airplane.

You're about to get airborne for the first time. That's what you're here for, but I've emphasized the procedures that come before the takeoff to make a point. Self-discipline is the name of the game. Personality and personal attitudes, not skill, are the crucial elements in safety. Here you can plainly see that the safety of your flights is up to you. Don't be ashamed to choose to be safe. The joy of being a pilot doesn't lie in a hundred thrills a minute, but in a gradual permanent change in your relationship with yourself and the rest of the world.

There is a story about a student watching an old geezer take twenty minutes to check out an airplane. The young pilot made some half-amused, half-scornful remark to his instructor and was quietly told he'd never be the pilot that old fellow was—Lindbergh was his name.

But enough; into the air!

4. *The First Lesson, Part Two—In the Air at Last*

Many people are amazed to find that they actually handle the controls—fly the airplane, in the air, themselves—on the very first lesson. Now that's going to happen to you! A last careful check for traffic in the air or on the runway, and the instructor lines up with the center line and opens the throttle. Don't just sit there, waiting for whatever wonderful thing is going to happen next. Look around, listen, and pay attention to what's going on. Look all the way out to the horizon in front of you and see if you can tell when the plane is off the ground. This has at least two advantages. First, it will help you to have that awareness of everything your senses are conveying to you that's so important; and second, it gives you something to do besides worry.

During the climb notice the slant of the wing tips relative to the horizon and see where the horizon is now in the windshield picture. While he's climbing, your instructor will show you what happens when he eases up on the right rudder pressure—the airplane rolls and turns to the left. Then he can show you that the right aileron, substituted for the rudder, can prevent that roll and turn, but you'll see that when he uses the aileron this way he loses airspeed and performance, so of course, the rudder is the correct control to use here. More about this later. Now listen to the sound of the normal climb to begin training your ears. And start right now looking for traffic in the air.

Learning to Look for Traffic

You can help yourself remember to look for traffic by imagining a little fantasy. Pretend that every other airplane in the air is flown by an enemy who's out to get you. Only by spotting him first can you save yourself. On a really busy Sunday afternoon this isn't so hard to imagine; in fact, you may begin to feel that it's true. I know it's hard to think about looking for traffic at the same time you're learning to control the airplane, but it's very important. A study of reports on mid-air collisions and near misses shows that an airplane in which a student is flying with an instructor is especially vulnerable.

I once tried out a device which could be set to ring a bell every minute or so to remind the pilot to look around outside. This will develop a reflex habit that persists after the bell is taken out, but I found it annoying during instruction. Another method that works well is to make a deal that the first one to spot each airplane seen during a lesson wins a nickel. With the amount of traffic today that could really add up.

Stability

After he brings the airplane to an appropriate altitude, probably above 2000 feet, your instructor will demonstrate each control separately. First, he'll show you how the elevator affects the aircraft attitude. He'll raise (or lower) the nose, and then he'll let go of the control wheel. He'll hold the plane straight with the rudder, and you'll see how the airplane noses up and down several times, but eventually returns, by itself, to a level attitude. This is one of the stabilities that are designed into the airplane—pitch stability. Mostly it depends on that downward force at the tail that I mentioned before. The demonstration will make clear that the airplane, if left alone, will not only correct but will temporarily *overcorrect* for attitude errors caused by misuse of the elevator control. You can see that the way to correct if you make this kind of error will be simply to restrain the airplane when it attempts to overcorrect.

The second control your instructor will demonstrate is the aileron. A very small aileron deflection produces a slow but perceptible rate of roll without any noticeable adverse yaw. In ordinary English that means that just a little aileron will make the airplane wing move down without any tendency for the nose to begin to turn the wrong way, that is, toward the higher wing. But extreme and abrupt deflection will definitely make the airplane start to turn the wrong way as the wing goes down. If the instructor establishes a bank and then rolls the airplane back to the wing-level attitude abruptly, without using rudder, you'll be able to see that the plane momentarily turns faster, as it is rolling back to level. And when the airplane is rolled abruptly without any rudder, notice the feel in the seat of your pants. You feel yourself thrown sideways, something like a skid in a car. Finally, your instructor will show you that the proper coordination of rudder with normal aileron feels smooth and prevents adverse yaw, and that even when the aileron is used to give an abnormally high rate of roll, enough rudder makes it feel OK.

You will see that the airplane has stability with regard to

bank too, but that it is different from pitch stability. In the demonstration of the elevator effect you saw that the airplane tended to return to a previously set attitude (set, that is, by the power-trim combination). But in the case of the wing we have a kind of passive stability (static stability). That is, though the airplane won't return to a wing-level attitude, it doesn't tend to continue to steepen the bank either. Actually this is only approximately true for most light planes, because, in fact, most will tend to roll, very very slowly, to the left when in a turn at normal cruise power. In a left turn the bank will gradually steepen, while in a right turn the bank will gradually become more shallow. However, this effect is usually so slight that you won't be able to see it consistently unless the turn is prolonged and the air is very smooth.

The Side-by-Side Effect

While you're being introduced to turns, pay special attention to the difference between the way a left turn looks and the way the right turn looks. The difference is there because you're not seated in the center of the plane, but on the left. When the plane is banked to the right (the right wing is low), your eyes are higher above the center line of the plane than they were when the wings were level, so the nose of the plane looks lower than it did before—you can see more ground over the top of the panel. This effect is well illustrated in figure 9–6 on page 52 of Kershner. But look at a spot on the windshield directly in front of you. (If it's too clean to find a spot, make one with a grease pencil.) Find one or make one that seems to rest on the horizon when you're in level flight with the wings level. You'll see that this spot also rests on the horizon in a turn in either direction. Just remember to look straight ahead of you, not over toward the center of the windshield, and you shouldn't have too much trouble with this.

A consistent tendency to climb in right turns and dive in left turns is a clue that you're having trouble with this effect. Sometimes commercial pilots, when they move to the right seat to give instruction after years of sitting on the left, have a problem with this. But their errors are the reverse of yours; they tend to dive in the right turns and climb in the left. These errors may be especially troublesome in the traffic pattern so try to correct the difficulty before you begin working on takeoffs and landings. You'll have to count on your instructor to watch for traffic while you're learning how the nose looks in turns. Once you have that down, make some turns and glance often at each wing tip. Especially begin to see how the wing tip looks against the ground when you're in a normal (20 to 30 degrees of bank) level turn.

The Rudder

After letting you work with the aileron, turning both ways and returning to level, the instructor will introduce you to the rudder. It is important to think of the rudder as a trimming device. That is, it is not used, like the elevator and aileron, to *change* attitude. It's normally used only to *prevent unwanted changes* in attitude or heading. But the instructor will show you that the rudder alone *can* be used to start a turn. And he'll show you that, once the turn itself has forced one wing down and the other up, and the rudder pressure is released, the turn feels just like any turn that was properly started with coordinated aileron and rudder. The elevator has to be coordinated properly with the bank during the turn, as Kershner explains, but the term "coordination" usually refers to the use of aileron and rudder together during entries into turns and recoveries from turns.

There is another important effect that the rudder may have: If rudder is used when a bank already exists, the nose attitude is affected. Rudder on the low side forces the nose lower, while rudder on the high side forces the nose higher. Thus a properly coordinated entry or recovery must include some movement of the elevator control too. Modern airplane design generally has reduced the need for rudder, though, and these effects may not be very easy to see when doing normal turns at cruising speeds. However, the effects are there. They will be more striking at slow speeds or when using more control deflection than usual, so they are factors in aircraft control during such crucial maneuvers as landings, or slow flight, or flight in turbulence.

Control Pressures and Tension

After each of these demonstrations you will handle the controls yourself. When you are operating the controls, pay special attention to how the controls feel now that there is flight pressure resisting your efforts to move the elevator, aileron, or rudder. Raise the nose to a new attitude and pay attention to what happens to that resisting pressure in the elevator. Notice that when you relax pressure, the controls themselves tend to return to a neutral position, the more so the longer you've held the plane in a nontrimmed attitude. But, even with the controls in the neutral position, the nose of the airplane will oscillate several times before it returns to level, as you saw in the demonstration of stability. Raise the nose again, and this time try to keep it in the new higher attitude. Notice first the light pressure it takes to raise the nose initially. Then pay attention to the way you have to move the control farther and farther back as the speed drops and how the nose seems to get heavier. And when you push the nose below the trimmed speed, you'll find increasing pressure will be necessary to hold the wheel forward as the speed builds up. The pressure you feel, which is partly the resistance of the air flowing around the control surface and partly the result of the balance of the airplane, feels different at low and high speeds.

While you're doing the flying, check yourself often for signs of tension. Make sure you're letting the weight of your legs rest on your heels on the floor. Tension can make you hold your legs off the floor using your thigh muscles, and if you let that happen you'll soon be exhausted. Remember to hold the control wheel lightly and to rest your toes lightly on the rudder pedals. Students and even licensed pilots often use so much pressure on *both* rudder pedals that most of the pressure they're fighting when they press on a rudder pedal is really the resistance, not of the air flow, but of their own foot on the other pedal. So when you press on a rudder pedal, make a point of lifting the toes of your other foot.

The Throttle as a Flight Control

There is one more control that directly and immediately affects the aircraft attitude. It is not thought of as a flight control and many pilots have never had its effects on the airplane's attitude clearly demonstrated, though most of us do learn eventually to compensate for these effects unconsciously. The control I'm talking about is the throttle. If your instructor skips this, ask him to go through it with you. Start with the power set at cruise and the airplane trimmed for level flight. When you add power, not touching any other control, the nose will go up slightly and the left wing will begin slowly to go down. Again starting from cruise, when you pull the throttle back, reducing power, the nose will go down. This is very noticeable, far more than the opposite, nose-up effect. As the speed builds up the right wing will slowly go down.

This set of reactions is normal in American-built airplanes. If you can't see all these effects in the plane you're flying, don't worry. Many training airplanes are out of rig or unbalanced in some way, or you may have an unbalanced fuel load. These conditions usually aren't dangerous—they only make the airplane's flight less efficient. In any case, it will be well for you to know just how the airplane does respond to power changes. I should point out here that even slightly bumpy air may make it hard or impossible to see these reactions clearly—another point in favor of smooth air for this so-important first lesson.

Straight and Level

After you've seen how each control moves the plane, and you know how they feel when you operate them, it will be time for you to try to hold the airplane in straight and level flight. Use a cloud in the sky or a mountain on the horizon to judge whether you're going straight. The reference should be far away and easy to see. Hold the nose in the attitude your instructor has shown you for level flight. While you work on level flight glance often at the altimeter, then look outside again and try to make gentle corrections in the nose attitude if you seem to need them. Hazy air, when there are no clouds and ground references are obscured, is really bad at this stage. You can use the compass, but you won't be learning the same things in terms of responses. The brain work necessary to interpret the compass will tend to keep you tense and block out all awareness of the feel and sound of the airplane. And if you depend on the compass, you aren't practicing the constant cross-check between the instruments and outside references that's so important in VFR flying, and you aren't developing the skill at seeing traffic that you'll need.

If conditions are good, and you are succeeding in staying pretty relaxed (unusual, but not impossible, on your first lesson), this is a good time for your instructor to start giving you some practice with the elevator trim. In the Piper Cub, where all the pressures were very light, I wasn't taught to use trim at all, but I have come to believe that it is important that trimming should become a reflex response to the awareness that pressure is needed on the wheel. In general, every time you move the throttle, you'll have to move the elevator, and eventually you'll have to use trim. And if you make much of a change in power and speed, you'll have to make a change in the rudder pressure too, at least for a while.

The pressure cue, which is what triggers the trimming response, depends on another, more basic, reflex response: the movement of your hand on the wheel to establish and *maintain* the desired attitude. Only if you *maintain* the desired attitude steadily through any power or speed changes can you judge the need for trim through the pressure of the wheel on your hand. If you allow the nose to pitch up or down, you mask the cues that let you trim correctly. With practice you'll develop some tiny reflex patterns—for example, hand pushes throttle, eyes check nose attitude, right toes exert pressure, left hand exerts forward pressure. All these adjustments are small, but if you don't make them when they're needed, you'll have larger corrections in attitude, altitude, and heading to make later.

Trimming and Relaxing

The ability to trim smoothly and easily depends on your being physically relaxed. A vicious circle can be set up here since your being relaxed will depend in large part on whether or not you are properly trimmed. So learning to trim properly is terribly important. If you have the control wheel locked in a death grip—with every muscle in your body tense and tight—you simply won't be able to feel the pressure that tells you that you need to trim, and when you do trim you won't be able to feel the reduction in pressure that tells you when you are properly trimmed. The sooner you learn to trim promptly and accurately, the easier the rest of your training will be. The only way to learn is through a lot of practice. Your instructor can start now by changing the trim slowly and slightly while you attempt to maintain steady level flight. When you are able to control the airplane's tendency to change pitch attitude in response to the trim changes he is making, you can start to practice returning the trim yourself to the normal position.

So far you've tried all the controls, flown straight and level, and worked with throttle and trim. The instructor's voice is giving out, and you may be just starting to feel comfortable. This is the time to turn for home. It's probably been about a half hour, or maybe a little longer. A little longer would be OK since so much of this lesson is demonstration. On the way back to the field you may have an opportunity for some more straight flying and a turn or two. The instructor will point out local landmarks, and when you arrive in the vicinity of the airport he'll demonstrate descents and leveling out from descents.

Another thing he may do on the way back to the airport is demonstrate hands-off flying. This is one of the most useful things you can learn and practice, especially to develop relaxed confidence, and it is too often omitted in flight training. If it isn't done now, it should certainly be part of your second lesson. Your instructor can show you that the nose attitude can be controlled by coordinated throttle and trim without touching the control wheel. It will be necessary to use rudder to control heading, and secondarily, wing attitude. With only a little practice you'll be able (in smooth air) to level out precisely at desired altitudes and establish precise airspeeds for climbs and descents. Later you'll find you can extend and retract flaps and make turns to precise headings too, all without touching the wheel. The two essentials are planning and making changes that are prompt but smooth and gentle. As you practice hands-off flying, you'll find out a lot about what is really happening when you fly the airplane. You'll gain a lot of confidence in the controllability of the airplane, and you'll be much more relaxed. If you practice this for a few minutes in each flight, your confidence in your own ability will grow and you'll make enormous progress in your awareness of the normal responses of the airplane.

Back to the Airport

The instructor will probably be talking during this lesson all the way to the ground. You won't remember or even hear everything he says, but it won't hurt you to have it there to listen to whenever you're in condition to absorb some of it. He will point out the landmarks he's using to find the field and to enter the pattern. He'll point out any traffic he sees as he checks the runway, the taxiway, and the pattern. He may be able to show you that a plane is easier to see when it banks for the turns around the pattern than when it's flying level on the straight legs. He'll describe what he's doing as he makes his approach.

When he turns the plane onto the final approach, look first at the runway and then look all the way out to the horizon and notice how everything seems to be happening much faster when you look at the ground closer to the plane. Remember to listen to the sound of the engine and the air over the airplane when the instructor first throttles back, and as he slows down and descends. When he's about to touch down, look out toward the horizon and to the sides ahead and try to judge when he will actually land.

The best part of the experience is over when you land, but

there is more you can learn. While you're taxying in, look around and work again on fixing the level attitude in your mind's eye. Watch your instructor as he goes through the final checklist and learn how to tie down the plane. It isn't "good operating practice" to walk away from an unsecured airplane, and it's a habit that can be costly. Generally it's not a good idea to leave parking brakes set. In some planes if they're left set for as long as overnight they may cease to hold, or if left set on a ramp in hot sun the brakes may be damaged.

While you're tying the airplane down, review the controls and their effects. If you have any new questions, ask them now. I keep talking about the importance of questions because most of us don't ask enough of them.

Before you leave the airport, the instructor will describe what he plans to cover during the next lesson. He'll assign appropriate chapters in Kershner or in whatever manual he uses, both for review and for preparation for the next lesson. He'll probably suggest that you get a manual for the airplane you're training in. And if he doesn't offer a copy of the checklists, ask for one or make your own.

Not Too Hard, Not Too Easy

Your future as a student and as a pilot may depend on how you feel about flying after this first lesson. Excitement and challenge should be there, but they shouldn't be overwhelming. It shouldn't look harder than it really is. Most people know perfectly well that they aren't supermen. If it looks to the new student as if it takes a superman to fly, he'll be discouraged very easily.

But you may be even worse off if it looks easier than it is. That can backfire in a couple of ways. I used to work for a man who was a natural pilot and a great salesman. He liked to take up groundlings and prospective airplane customers and ostentatiously fly around with only one finger on the controls. Like the manufacturer's ads that talk about driving in the sky, he made flying look very easy. And he did sell a lot of airplanes to beginners, and he sold the course of instruction to go with them. But later, when the proud new owner noticed how hard he was working to learn this supposedly easy thing, his ego was injured. He felt stupid because learning to fly wasn't easy for him. Then he would get the idea that, after all, it did take a superman, or at least a special knack or talent, and that all this was just not for him. We lost potential pilots that way, which was bad enough. But the other way it can backfire is more serious because it affects safety in the air.

What may happen is that the pilot-to-be, believing that flying safely is easy to learn and easy to do, understandably refuses to take the trouble to learn the skills, and develop the judgment, that really are required to make flying safe. He may squeak through his flight training, but it will take a change in his attitude to make him a safe pilot. This may happen if he's lucky enough to survive one of those hair-raising, I-learned-about-flying-from-that experiences, but that's the hard way. So don't let yourself get the idea that it's all going to be easy.

One more thing: Think about the relationship that's developing between you and your instructor. By the time this lesson is over, especially if it's actually the second time you've seen each other (having had a preliminary interview before scheduling this first lesson), both you and the instructor have formed some opinion of each other. You should have made a start on developing mutual trust and respect. If you don't feel this is happening, think about the reasons, and give it a little more time. Maybe that whiskey voice comes from his long day of hard instruction, and his big red nose is just as innocent, while that clean-cut, articulate, well-dressed young instructor you wish you were flying with is actually running dope on the side. If your instructor was late, or impatient, or preoccupied, remember this could be the exception for him, so give him another chance. But be forewarned; if you have an exalted image of him, he won't go out of his way to disillusion you. He knows you'll pay more attention to what he says if you think he's God. And that's all right, if it motivates you to do your best. But don't let it keep you from asking questions. Ultimately, you'll have to develop your own standards and goals and become independent of your instructor no matter how Godlike he appears.

An artist I know who does a lot of teaching put his finger on the real crux of the matter when he said, of teaching, "It's a problem in human relations—not between the teacher and the student, but between the student and himself." Finally, it will be yourself that you'll have to satisfy.

5. *Basic Airwork*

For me it took only one lesson and I was hooked—badly hooked. I thought all week about what I had seen and how it had felt. I got edgy about it: Could I really do it? How was the throttle going to feel? (My first lesson was in rough air and all I did was try to keep the nose and wings level with the stick.) Would I ever be able to fly alone? Flying was like a real addiction for me for years. I felt great—euphoric—immediately after a fix, I mean flight. The glow lasted a day or two. Then it began to wear off. I started doubting my ability—sure, I had done it before, but maybe that was just luck. Could I really do it again? I got jumpy. Getting into the airplane again, I was tense and nervous. Finally, I was in the air, climbing, and a big grin would spread across my face and the inner glow was rekindled. I knew I could do it.

More About Helping Yourself

Caring about it that much will obviously help you learn (provided the doubts and tension aren't extreme). But even if you aren't hooked that hard, review that first lesson in your mind. Yes, I know I've said this before, but it's important enough to say again. Go over the description of it here. And read Kershner again too. If the experience of the flight is blurred in your memory, the review will bring it into focus, and even if you think you remember it all, the review may remind you of something, or raise a question you'll want to try to answer when you fly the next lesson.

Right here is the place to say that when I talk about first lesson, second lesson, etc., I am referring only to a general sequence. It could easily require three half-hour lessons to cover the demonstrations and practice I described in the last chapter. So if you arrive at the point of my "second lesson" with a couple of hours already behind you, don't worry about it.

The permanent base, the foundation, of all your future flying is laid in these early hours. Bad habits formed now will be tough to break, and they'll cause you more and more trouble. Good habits are easiest to form now. If you're already beyond this early stage and have some bad habits, it's not too late; it will just be a little harder. There are some bad habits that almost everyone has trouble with and that most instructors don't seem to deal with effectively. The one that always comes to my mind first is the matter of turning the wheel when what's really wanted is a little rudder.

Dealing with Wheel-Steering

Believe me, you'll do it. We all do it, or at least we have done it. What cured me was an overabundance of pride. The Cub has a stick, of course, so learning in the Cub I skipped over that problem. You would think—at least, I thought—that a person who has learned, in a Cub, that it is the rudder that turns the nose could climb into a newer airplane like the Cessna 172 and not fall into the trap that the design of the wheel has laid for us. Well, 'tain't so! My instructor had told me how people tried to steer with the wheel and I had allowed as how I would never do that. One day he had an errand to do in a Cessna 172. He let me come along, and after doing the errand at a nearby field, we headed back. As we taxied out for takeoff, he let me handle the controls of the 172 for the first time. I was thrilled and was concentrating hard on being perfect: center of the runway, smooth throttle, etc. We came to the jug handle at the end of the hard surface, and I turned the wheel just like all those people I had been so scornful of. My instructor burst out laughing (imagine how it hurts to have God laughing at you), and he was still chuckling when we got home. And, of course, he told everyone this good joke on me. I deserved it, but I made up my mind it would never happen again and it didn't.

There is another way to cure this habit that works for some students—one I've never tried myself. I found out about it when I flew with a low-time student in a Cessna 140. I was amazed to see that he had no tendency to turn the wheel while taxying. I had only recently started instructing in a plane with a wheel and I was finding it a real problem for my students, so I asked him how he had learned not to use it like a steering wheel. He said he couldn't tell me—he wasn't willing to repeat the language his instructor had abused him with.

I've flown with pilots with several hundred hours, even instructors, who still move the wheel that way. The problem is that this is more than just another bad habit; this is a reflex and it takes more than just deciding not to do it to actually stop. The slightest motion of the wheel during straight taxying is a giveaway that your wrong reflex is in there working. And every time you let it happen, you strengthen the reflex pattern and make it harder to learn not to do it. What I'm leading up to is an urgent plea to take advantage of all your taxi time to *practice not steering* with the wheel. It won't be easy. You have to keep your attention on two things at once —*using* your feet to control direction and *not using* your hand—but it's worth it to unlearn this wrong response early in your training.

You might think that just keeping your hand off the wheel would break the habit. A lot of instructors recommend that, but it simply doesn't work. As soon as you put your hand on the wheel again you'll be trying to steer with it—your reflex will be just as strong as ever; your hand hasn't unlearned it. So keep your hand on the wheel but force yourself not to move it *at all* during straight taxying.

The reason I say "straight taxying" is that you actually should move the wheel most of the time when you make a turn on the ground. This isn't essential if there is no wind, or a light wind, or if you're flying a low-winged plane even in strong wind. But there are some planes in which it can be very important. And since you're going to be thinking about the wheel anyway, you might as well learn this now, even though you won't be out in the kind of strong winds that make it critical for a while yet.

Taxying in the Wind

The idea is to use the aileron and elevator *on the ground* to keep the wind from getting under a wing or under the tail and upsetting the airplane. The rules for the position of the aileron and elevator are simple: If the wind is coming from any point ahead of the wing tips, use the aileron the same as you would use it in the air to keep the windward wing down, and hold the elevator slightly aft. If the wind is coming from a point behind the wing tips, use the aileron opposite to the way you would use it in the air, and hold the elevator forward. The hard part is keeping track of just where the wind is coming from in relation to the wing tips and the nose, especially as you turn.

Don't think all this isn't important. Failure to handle it properly has caused some bitter humiliation to some highly experienced pilots. The Tri-pacer was an airplane that used to cause special problems because there was a linkage between the rudder and the wheel. It was there to make coordination easier and to "spin-proof" the plane, but it meant that you were set up for trouble when you taxied downwind and then started a turn. That is, the aileron went the wrong way unless the pilot used some effort to override the spring linkage. One airline-rated pilot flipped over in a mere 22-knot wind in a Tri-pacer in full view of the control tower. Imagine his feelings! Wind, airplanes, and the facts of aerodynamics are no respecters of ratings or human dignity.

You can get double duty from your taxi practice by using the wheel carefully *only* and *always* as it should be used. If there isn't any wind at all, pretend the wind is straight down the runway. Review all this with your instructor. Study the pictures in Kershner, figure 6-5, page 38. Plan your turns and what you're going to be doing with the aileron and elevator. It will help a lot if your instructor watches you very closely and lets you know every time he catches your hand trying to steer, but it will help even more if you catch yourself.

Looking for Traffic

Another bad habit is looking straight ahead. A lot of emphasis on looking for traffic will help here, but looking straight ahead is a habit a lot of instructors don't notice in their students, especially new instructors. When I teach would-be instructors, I play student, and I always do some flying—usually straight and level—in which the only error I'm making is that I'm looking straight ahead—not looking around for traffic at all. I can't keep it up very long; it makes me too nervous. But I've never had a flight instructor applicant yet who could figure out what error I was demonstrating.

Most of the time you're in the air you'll be flying straight and level, so most of the time you should be looking for traffic at your own altitude. But near the airport, where the traffic is most dense, the greatest hazard may come from above or below. Experience will show you that you can see airplanes more easily on cloudy or moderately hazy days than you can when the air is perfectly clear. I have already mentioned that you see airplanes more easily when they bank and turn. Some pilots I know make a practice of rocking the wings periodically when they are flying cross-country, just to make themselves easier to see. One of the very nicest things about the flight I made across the Atlantic in a light plane was the deep certainty I felt that no one was going to run into us—there just wasn't anyone else out there!

Climbs and Glides

During your second lesson you'll be practicing climbs and glides. You'll get the best practice if you make extended climbs and glides, over at least 2000 feet of altitude, leveling out every 500 feet. The big advantage to doing it that way is the amount of practice in trimming that you'll get. Remember to listen to both airframe and engine sounds and to note how the nose looks and how the wing tips slant across the horizontal. When you level out, don't watch the altimeter—watch the nose attitude. Only check the altimeter to see if the attitude you're using is working to hold the altitude. Some experience and some help from your instructor will show you just how early to start leveling out from the climb and the descent.

Unfortunately, this is not a simple matter of learning *the* climb and *the* glide. There are four principal types of climbs. In order of increasing airspeed they are: best angle of climb (achieved at a speed which is designated V_x; best rate of climb (this speed is designated V_y—you can remember that V_x is best angle because an X has more angles than a Y); recommended climb (usually a higher speed than V_x or V_y for better engine cooling, but still giving good climb performance); and cruise climb (a more gradual climb at cruise speed or close to it). In light, training-type planes, full power is desirable for at least the first three of these climbs. Backing off the throttle a little invites overheating, because a small portion of unburned fuel in the overrich, full-throttle position helps to cool the engine. During this early training you'll probably use only one of these climbs and it will probably be the recommended climb. But get that straight with your instructor.

The various descents are even more confusing. There's a power-off glide and a power-on glide, each with or without various degrees of flap extension. There's best glide speed (usually about the same as V_y—best rate of climb speed), which will give you the most distance over the ground for your altitude. And there's normal glide speed, now considered to be 1.5 times the power-off stalling speed (designated V_{so}), which will usually bring you down a little more steeply than the best glide speed. Make sure that you and your instructor are talking about the same thing when you use the word "glide." To me it ought to mean a power-off descent, but the word is used more loosely now.

Back in the days when the Cub and the Aeronca were the training planes everyone learned in, a normal glide was always power-off, and we didn't have the option of flaps to confuse us. There was only one power setting—engine idled—so there was only one engine sound. There was only one airspeed, so there was only one overall engine and airframe sound to learn to recognize. Since we had no flaps, there was only one attitude which was right. You can see that it must have been easier to learn to glide properly. Some instructors solve the problem of variation in the descents by not using flaps in initial training. That doesn't seem appropriate to me.

Usually when you learn the fundamentals you learn a power-off glide. But in today's typical airport traffic situation you won't be able to make most of your approaches power-off, so it is probably more practical to learn a glide or descent using partial power. You'll have more time in the glide to look around and listen, and you'll be treating the engine better too. You will find that you can learn to separate the sound of the air flowing around the airplane from the sound of the engine. Even using partial power, which may mean slightly different engine sounds on each glide, you'll be able to learn to recognize a normal glide by the sound.

Instruments for Feedback

While you're devoting some real attention to how things look and sound in climbs and glides, there's an exercise you can do that will give you good practice in judging speed. Prolong the climb or descent and alter the airspeed to ten miles higher and ten miles lower than normal by changing the nose attitude. Hold the new nose attitude long enough so the airspeed becomes stable at the new figure. Notice how long it takes the airplane to slow down or speed up *after* you've changed the attitude the right amount. Notice how much difference there is in the sound; not much, but it is perceptible. You'll be able to use the airspeed indicator to give you effective feedback about your nose attitude after you've worked with this a few times. The lag in the airspeed that you'll hear people talking about is mostly not in the airspeed indicator. While there may be some lag there, most of the lag is the time it takes the airplane to accelerate or decelerate to a new steady speed when the attitude is changed.

As you begin to use the instruments in the second lesson and the following ones, you'll probably notice a strong temptation to rely on them. Remember that if you become dependent on instrument readings you'll be missing a lot that's important. The current emphasis on how becoming instrument rated makes you a better pilot can be very misleading. The result ultimately may be a lot of very precise IFR (Instrument Flight Rules) pilots who have lost or never developed many of the most important skills of the VFR (Visual Flight Rules) pilot. For example, can they make turns through approximately the desired number of degrees without reference to the compass? The ability to correctly estimate angles is an important skill to develop, because turns are a basic part of the traffic pattern where there is so much else to be thinking about. When you're asked to turn 90 or 180 degrees, try to make the turn by reference to the ground. Check the directional gyro before you start and after you roll out of the turn. If your turn wasn't just what you wanted, file away the error in your head, and try again. If you have a lot of trouble learning to judge those angles, start by practicing across a straight road, or a railroad. Most of the clues to what is going on will be there for the seeing if you're looking outside, but you won't learn how to see them if you're concentrating on the compass.

As you practice climbs and glides to specific altitudes, judge the speeds on your own, only looking at the airspeed indicator to correct your judgment. Remember to use sound and the feel of the controls as well as how the nose looks. Try slight rolls left and right at varying speeds and see if you can begin to feel differences in the flight pressures and in the response of the airplane at speeds above and below normal cruise. Start trying to judge altitude, too, at least while you're in the pattern. Use the instruments mostly for feedback to help you learn how to interpret accurately what you see and hear in terms of performance. Use the instruments, in other words, to develop the skill that will give you well-founded confidence in your own judgment.

Flaps

You'll be introduced to flaps sometime during this period. You'll be shown how fully extended flaps in level flight hold the speed below cruise even with full power. You'll see how the glide angle is steepened if a constant airspeed is maintained as the flaps are extended. And your instructor will show you how much the flaps affect the landing (and takeoff) distance. In effect, flaps are a device to change the design of the wings temporarily, providing a high-lift, low-speed wing for takeoffs and landings without detracting from cruise performance. Experiment with the flaps in the air. Find out whether the nose pitches up or down when flaps are extended and retracted, and whether the tendency to pitch is the same across the whole range of flap speed and at all degrees of flap extension. Find out whether flaps change the rudder pressure needed to hold the airplane steady at various combinations of speed and power. In some airplanes with electric flaps you can learn to judge the degree of flap extension by being aware of the feel of the plane as the flaps change. In particular, in the Cessna 150 and 172 this means that you can set your flaps at 10, 20, or 30 degrees during the landing approach without looking at the flap indicator—a good trick because you're free then to be looking outside during every possible second of the time you spend in the airport traffic pattern.

Pattern and Track Maneuvers

So far, you've been dealing only with ordinary flying—the basic stuff that you do in the air to get from A to B on a nice day. You haven't had anything to distract your attention, in the smallest degree, from controlling the airplane—at least, not anything other than your own natural apprehension, and such anguished screams as your instructor has been unable to suppress. The next step may be slow flight and stalls (and spins if your instructor is so minded); or it may be pattern and track maneuvers, which are composed simply of more straight and level and turns, but closer to the ground and with the added pressure of trying to fly a particular track over the ground. Which of these lessons you get next may depend on the weather. If the visibility is poor, the ground reference maneuvers will be better because you'll be low enough to have a good horizon anyway. On the other hand, if there's a lot of wind and associated rough air which smooths out at three or four thousand feet, then the slow flight and stalls will work out better.

If weather isn't a factor you'll benefit more from a session on pattern and track maneuvers. The obvious reason for doing these maneuvers is to learn to recognize wind drift and to correct for it effectively, but there is another advantage to being introduced to ground reference maneuvers at this point. As you try to follow a winding road, or fly S-turns across a road, or fly a rectangle around a field, your instructor can get some idea about how fast you're developing the reflex patterns which are so much of flying. Do you coordinate rudder and aileron even when you're concentrating on getting the nose around to cross the road at a right angle? Are you aware of high or low airspeed if you mishandle the elevator during and after turns? Have you begun to have confidence in your own judgment? For example, if you hear a high or a low airspeed, do you look first at the indicator, or do you look first, as you should, at the nose attitude? Have you begun to develop some ability to judge altitude when you're within 1000 feet of the ground? What your instructor observes during this lesson will help him decide whether you are ready to go on now to an introduction to slow flight and stalls and then begin takeoffs and landings, or whether, instead, you need some more practice on the fundamentals.

Be sure you understand slipping and crabbing before you go up for this lesson. Slips are caused by nonstandard use of the controls; they are inefficient so they're only used when some special condition outweighs performance considerations, most often on crosswind approaches. Crabs, which may also result in movement over the ground toward a point which is not straight in front of you, are nothing more or less than normal straight and level flying, but in air which is moving over the ground in a different direction from the direction you're headed.

In all ground reference maneuvers you're simply trying to fly so that the airplane stays over some particular line on the ground. The line may be a real one—a road or stream—and that's tough enough in a stiff wind. When the line is an imaginary one as in the S-turn across a road, or one of the eights, or the turn about a point, the problem is magnified. Most people do a poor job on these unless, consciously or unconsciously, they imagine the line they are trying to follow. And that isn't easy either, but the traffic pattern where you'll apply what you learn here offers the same problems and you must learn to solve them.

➞ *S-TURNS.* S-turns are often considered the easiest but they offer several difficulties even aside from the problems of correcting for wind. First, a straight section of road is hard to find in some parts of the country. Second, the references you use to help you imagine the loops must be changed constantly. You may be following the track you think you want, but you may be misjudging the size or shape of your loops, and so fail to fly the desired half-circles on each side of the road. Finally, if you have a strong wind across the road, it will be virtually impossible to roll from the steep turn in one direction into the steep turn in the other direction fast enough to make the loops even and of a reasonable size.

➞ *RECTANGULAR COURSES.* The rectangular course is a particularly nice track maneuver, but has some problems too. It may be hard to find a good rectangle on the ground. The one that used to be my favorite—a really nice one with big fields for forced landings and even a private airstrip along one side—is now covered by the spreading edges of a huge development. The next good area I found—flat, out in the country, big fields—was spoiled when they built a tower which rises to 1049 feet right on one side of it. While it is barely possible to practice this maneuver by flying some distance to one side of and parallel to a straight stretch of highway, turning and flying straight across it, then turning and flying back on the other side of the highway, it's not effective in learning to turn onto specific desired tracks. So it is worth really looking around the countryside for a good rectangle.

If you need extra practice on climbs and glides and on climbing and gliding turns, you can add them to the basic rectangle. And if you have trouble remembering all the procedures you have to squeeze into the pattern—if the pressure of other traffic and the excitement of making takeoffs and landings make your mind go blank, and you find yourself forgetting to throttle back when you start your approach—you can practice all these steps while you fly a rectangular course up in the air, simulating a traffic pattern without the pressures inherent in the real thing. Then when you have the procedures down cold, you can come back into the pattern and handle it more comfortably.

➼ *EIGHTS.* You probably won't do eights yet but I'll include some comments on them here for convenience. When I learned to fly, *the* track maneuver that we did on a private flight test was the eight-around-pylons. That was dropped in favor of the 720 turn about a point, and the explanation I heard was that it was too hard to find two points close to each other, both of which could be used without upsetting anyone on the ground. I can't imagine that that is any less true today, but the latest changes in the certification procedure include eights among the list of maneuvers which the applicant may be asked to demonstrate.

Eights are more demanding than circles because the pilot must estimate his drift and plan the transition from one turn to the other so as to compensate for the wind and keep his loops even. This offers special problems because you will normally be flying close enough to the center-point reference so you won't be able to see the reference for the right-hand turn until after you get the wing down and start the turn. The best way to deal with this difficulty is to use the center of a crossroad for the center of the right turn, using a telephone pole or a tree along one of the roads for the other center point. You'll be able to see distant sections of the roads and so can estimate where the crossroad is in order to time the start of your right turn.

➼ *720 TURNS.* The 720 turn about a point has been a problem ever since it appeared in the flight test guide. First, it is confused, by a lot of pilots, with the 720 power turn, or steep 720—now no longer part of the private flight test. I flew with one instructor who had somehow gotten all the way through the certification without ever getting that straight.

His students all had a terrible time with the 720 about a point, so I went out to fly with him one very windy day. His performance was amazingly awful! He used violent slips and skids; but because his track was approximately circular and his bank was approximately a constant 45 degrees, he thought he was doing well. He simply didn't feel the horrifying things he was doing, and because the ball on the turn and bank indicator was jumping around due to the rough air, he was lost. He was a victim of the current training practices, especially the emphasis on the ball to determine whether or not the use of controls is coordinated.

He can hardly be blamed for that. He was the product of an instructor, who was the product of an instructor, who was the product of an instructor, who had never learned to feel a slip or a skid. In the old days most trainers didn't have a ball to depend on. The pilot had to learn to feel proper coordination. That meant that the instructor had to teach him. So one of the things that used to be done on every instructor flight test, and a thing for which the applicant could be and would be failed, was a check of his ability to pick up and diagnose errors in coordination. The inspector would fly a turn with a slight slip or skid. The would-be instructor had to sense it and correct it. Sometime since then this has dropped out of the flight instructor flight test. The result is that most instructors today don't know that a pilot is slipping or skidding unless they can see the ball, and if it's a heavily damped ball, like most of the newer ones, the pilots they turn out are often poorly coordinated, and have little feel for the airplane. A 720 in a stiff wind is likely to show this failing.

The notion that this maneuver should be flown in a constant bank is relatively rare. A more common misconception is that the wing should point constantly toward the center of the circle. I flew once with an unusually well-trained and skillful applicant for the private certificate. I was really enjoying riding with him, when to my surprise his flying suddenly went to pieces as he attempted the 720. I found that he thought the wing was supposed to point toward the center at all times throughout the turn. To make that work while flying a circle he had to slip and skid, so he did. I asked him if he hadn't seen the wing move forward and back relative to the center when his instructor had demonstrated it to him. Yes, he had, but he'd thought the instructor was making a mistake and he didn't like to question it. If he had, he would have learned that if there is any wind at all, the wing points directly at the reference point only twice in one full circle—when the plane is headed directly upwind with the shallowest bank, and when it is headed directly downwind with the steepest bank. The rest of the time the wing should point behind or ahead of the reference. If the usual error is made, being too late in changing the bank, the shallowest bank as well as the steepest bank will come a little late, while you're correcting for the error.

On the flight test in question I gave a quick demonstration and the applicant, who flew unusually well, was able to do the turns like a pro. It is often such misconceptions that are the root of difficulties students have with maneuvers. In this case, while no one had ever told him that the wing should hold directly over the point, no one had ever told him that it shouldn't either. A careful briefing or a question from him would have solved the problem.

Understanding this maneuver will have particular value in reference to the final turn when there is a crosswind. Along with developing a clear understanding of how ground speed affects the bank required, do some thinking about how that ground speed affects the *time* you have available to see what's happening and correct for it. This is a significant factor, especially when you're in the pattern with a wind that's behind you on the base leg. I've seen commercial pilots with hundreds of hours overshoot the final turn several times in a row under this condition.

If your problem seems to be mostly a matter of not being able to see that imaginary circular line you want to fly over, you may find it easier to fly a circle that remains inside the limits of a square field. Once you've done that a few times, you should be able to use a center point more easily.

One point about practicing the various pattern and track maneuvers: Give careful consideration to the area over which you do them. It is much easier to see your track when you are as low as possible. But you have to think about the people on the ground, and 500 feet AGL (Above Ground Level) is the minimum distance from a "person, vessel, vehicle, or structure." Even a fence post counts as a structure. And if you're going to be this low you better be over an open area where you could land if the engine quit. It just doesn't make good sense to go out over the New Jersey Pine Barrens, or the Great Dismal Swamp, or some other thoroughly unsuitable (to say the least) emergency landing area and then do low airwork. If that's the only choice you have, OK. Engine failures are rare—a lot less likely to happen than a blowout or a brake failure in your car, for example. But try hard to find a place where you could make a successful (successful here means without damage to the airplane) forced landing if you had to. Maybe you can find someone in your area with a private strip who wouldn't mind your practicing overhead. While you're doing these maneuvers, get familiar with how much room you need over the ground for a normal turn. You'll find this knowledge will really help when you start doing takeoffs and landings.

There is a general tendency to look on all these maneuvers as test maneuvers rather than as an integral part of the complex of flying skills, although it is obvious that they have some application to the traffic pattern. The point is that a really skillful pilot—one who can fly an airplane with unconscious timing and coordination—could do any or all of these maneuvers the first time he was shown them. Such pilots are few and far between, but this level of skill is the goal we all are striving for. These maneuvers, then, are a way of practicing good smooth control of the airplane while you meet the demands of sensing and reacting to the facts of flight. You'll find a lot of pleasure in doing this practice in the really smooth air you can find early in the morning. And, flying alone at the hour when alarm clocks are ringing in the houses below you, you'll feel as if the world belongs to you.

6. *Slow Flight, Stalls, and Spins*

Stall demonstrations and practice have been a very important part of learning to fly almost since the beginning of aviation. Today's airplanes are mostly very safe to stall. By that I mean that they will not go into spins unless they are badly mishandled. But, once in a spin, some of our modern airplanes may be harder to bring out than the older, less sophisticated airplanes. And those aerodynamically naive slowpokes crashed differently too. In olden days when the world was young, an unscathed aviator might emerge from the twisted pile of shattered spruce and torn fabric. No more! Today's metal and plastic marvels, built for speed and packed with electronics and Naugahyde, hit a lot harder!

The Spin Controversy

There has been, at least since I learned to fly and probably long before that, a continuing controversy about spin training. It was not required for private or commercial pilots at the time that I was a student, but instructors did have to demonstrate controlled spins left and right when they were tested by the FAA. There have been many movements to try to put spins back into the curriculum for nonprofessional pilots. I disagree strongly with the reasoning behind this effort, even though I have great admiration and respect for many of the pilots who advocate the return of spins.

I figure it this way. An airplane won't spin until it stalls (or at least until one wing stalls). And, except in considerable turbulence, very steep banks, or violent use of the controls, an airplane won't stall until *after* it is flying much more slowly than is normal. Furthermore, no amount of practice in spin recoveries can make a really substantial dent in the stall-spin accident figures because so many accidental spins are entered at altitudes so low that no amount of skill could alter the outcome. I do agree that the matter needs attention, but I don't think spin training is the answer.

What Comes Before the Spin

To begin with, I'd like to see more attention paid to the circumstances that lead a pilot to attempt slow flight close to the ground, slow flight that sometimes results in inadvertent stalls and spins. There are two situations that are particularly frequent preliminaries to fatal accidents of this kind. Both situations occur in the traffic pattern around the field.

Let's take the slow approach first. This may occur either because the airplane is low and the pilot unconsciously raises the nose, or it may occur because the pilot is trying to maintain spacing behind other, perhaps slower, traffic. I believe it is a habit of overdependence on the airspeed indicator that is the basic factor in these accidents. That may sound paradoxical, and I don't mean that the airspeed indicators have malfunctioned to cause these accidents. I mean that the pilots have never learned to be aware of the clear indications the airplanes give when they are starting to get too slow—sloppy controls, poor aileron response, the need for more rudder with the extra aileron that's necessary to be effective at slow speeds. All those signals are certainly being sent, but they just don't reach the pilot's awareness because he hasn't been schooled to attend to them. The airspeed indicator is sending a signal too, but it isn't going to help him, because he isn't looking at it. He's looking outside, where he really should be looking—outside where the traffic, the runway, the obstacles are. All the training on using a good inside-outside scan won't keep him from fixing his eyes on the plane he's overtaking or the trees that seem to be rising between him and the runway. Good early practice in hearing, seeing, and feeling how the airplane behaves in slow flight is the best protection you can get from a stall that could occur at low altitude, when your attention is diverted from the instruments.

Getting Over the Fear of Stalls

If you are one of those, and there are many, who are so apprehensive about stalls that you hardly know what's happening—you just grit your teeth, shut down your perceptions, and pray until each stall is over—your practice isn't going to do you much good. Theory says the fear will lessen with familiarity, but I'm not so sure that's true. And I am sure there are many people who learn mechanically to do excellent stall demonstrations and recoveries but never lose that almost-but-not-quite-paralyzing fear.

There is a way of working into stalls from slow flight which may demystify them and give you a lot of good practice in being aware of those vitally important sight, sound, and control response cues. Go up and trim the airplane for flight at a power setting that will just permit very shallow level turns without stalling. Now, with occasional turns for monitoring traffic, raise the nose gently to the point where you are on the edge of the stall. Lower the nose to regain flying speed. Contrary to normal stall practice, don't add power; that will complicate matters and add confusion. The point now is to get comfortable with this slow flight. Do it again and again. If you're doing this at the right power setting, each sequence will result in a small loss of altitude, so you may have to break off and climb periodically. Continue to do this, both with and without flaps, and keep on doing it until it isn't scary any more—until it becomes boring. Then start holding the elevator back a bit longer, trying to hold the airplane perfectly straight, and the wings level, by fanning the rudder. Again, do it over and over until it's boring. Then start adding power at the point where you can fly out of the almost-stalled condition without losing altitude. Notice the right rudder you need when you add power. Try this practice in turns too. After all this is boring, do it all again, but with more power initially, which will make the stall more distinct.

Some might argue that the development of a reflex response—addition of power along with relaxed back pressure—is the most important goal, and that this practice won't contribute much to that. I agree that that's a desirable goal, but the fact is that very few pilots do enough stalls to make that response a strong and sure reflex. So I believe that it is more important to be able to recognize the clues to borderline airspeed and to learn to use the controls properly to maintain a wings-level attitude while on the edge of a stall. This practice will contribute to both these goals while it helps the timid pilot lose his debilitating fear of stalls.

The Stall-Spin in a Go-Around

The other situation in the pattern which seems to precede a lot of stall-spin accidents is the go-around which is initiated when the airplane is almost at landing speed. This stall is more of a problem in some airplanes than in others (as are stalls themselves, for that matter). The key factor is the amount of yaw caused by full power at low speed with flaps extended, and also important is the adverse yaw caused when full aileron is applied. If there isn't more yaw than the rudder can handle, you can recover from this stall without much trouble, provided you have enough altitude. But some airplanes just don't have enough rudder to handle mistakes that may be made in this situation.

Let's look at what happens in a Cessna with full flaps when there is a sudden application of full power at an airspeed just above stall. There is an immediate strong tendency for the nose to swing to the left due to power effect. Full right rudder will only barely control that tendency. When the airplane is trimmed for the landing, there is also a tendency for the nose to come up, which will slow the airplane despite the power and make the stall more likely. So far, the unpleasant results can be prevented by proper use of the controls—right rudder and forward elevator. But most people don't respond immediately with rudder to control this power-effect yaw. Instead they use the aileron because of that wheel-steering reflex. The aileron added at this point causes adverse yaw which combines with the power-effect yaw to *increase* the unwanted turn. There just isn't enough rudder, even if full rudder is now belatedly applied, to correct for both the adverse yaw and the power-effect yaw. The result may be a spin entry.

There are two preventatives for this. First, start your go-arounds early. Second, learn to expect the yaw that comes with power application and be ready with whatever rudder pressure is necessary. When you check out in a new airplane, find out about this in your initial check-out.

The go-around stall, with full power and full flaps, is not one which the FAA includes among those to be demonstrated by the private pilot. I've heard an FAA inspector say these stalls were too tough and dangerous to expect a private pilot applicant to demonstrate them. But the airplane and the laws of aerodynamics won't show compassion for the low-time pilot any more than they will respect the dignity of the high-time pilot. This stall does happen and can be a killer, so do learn how to prevent it.

The Cross-Control Stall

The cross-control stall is another that used to be listed as one of the variations that the private pilot had to be familiar with. It is one that can lead to a spin at low altitude and you ought to see what it's like, even though it's not likely to occur in a modern airplane. We used to do them from both slips and skids, and it wouldn't hurt to see both kinds, but it is the stall in a skidded turn which presents the worse hazard, and there's a not uncommon situation close to the ground that can trap the unwary pilot in a series of actions which result in just this kind of skidded turn.

It can happen like this. On base you start your turn a little late, maybe because you haven't allowed for the effect of the crosswind behind you. Seeing your error, you steepen the bank, trying to get around the turn in time to be lined up with the runway. Your airspeed may be a little low because the wind behind you on the base gave you a visual illusion of higher-than-normal approach speed. Now the wing looks uncomfortably close to the ground (to your unconscious), and besides you may have heard that you should avoid steep banks in a glide (not necessarily true). So you press a little inside rudder to get around the turn without steepening the bank. That rudder on the low side tends to steepen the bank anyway, of course, as well as forcing the nose lower, so you respond by using the opposite aileron and pulling back on the elevator. This is exactly how we do a cross-control stall from a skid. When this stall occurs, unlike a normal turning stall, the low wing suddenly whips even lower, the nose drops, and you find yourself looking straight down at the ground. In the split seconds during the stall entry you'll probably do everything wrong—that is, you'll use even more aileron in an attempt to stop the roll, and more back elevator. The result can be a full-fledged spin. Plenty of pattern and track practice developing unconscious coordination is the preventative for this one.

The Accelerated Stall

The first problem with the accelerated stall is its name. It invites confusion. By definition any stall that occurs above the "normal" airspeed is an accelerated stall, and it is always caused by a load on the wing that exceeds the maximum gross weight. Now, there are four ways to get an increase in load. The most obvious is simply by overloading the airplane. By definition, any stall done while you are overloaded will be an accelerated stall. A second cause is turbulence. Gusts in rough air can, and often do, cause momentary accelerated stalls. As a matter of fact, maneuvering speed—the speed recommended in rough air—is actually a stall speed. It is the stall speed for the airplane when the wings are carrying the maximum allowable load. Once stalled, the wings are no longer bearing a load. Thus, if you don't exceed maneuvering speed, you cannot exceed the design load limits of the plane, risking damage and loss of control, because you will stall first.

The other two types of accelerated stalls are those that

have been called for on flight tests at one time or another. They are 1) those caused by the load induced in a steep level turn, and 2) those resulting from abrupt use of the elevator control. When you attempt to practice these, it is often so difficult to get a stall that you will come down from your flight believing that this kind of stall just couldn't happen in real life. Not true. They do happen, and they do cause fatal accidents on a shockingly regular basis. Let's try to see how that can be, so we can see how to avoid them.

There are two situations that are most often associated with accelerated stalls in turns. One is the turn made around something, or more often someone, on the ground. In the effort to stay in close and keep the area or person in sight, the pilot gets too slow, with too steep a bank. He's likely to combine that with a slip in a low-winged plane, or a skid in a high-winged plane, in order to keep the wing out of the way of what he's trying to see. Either one can only make matters worse. Notice, incidentally, that the pilot's attention is again necessarily outside, not on the instruments. If the stall is not also a cross-control stall, the accelerated stall from a turn should be easy to recover from. For one thing, since the airspeed is above the normal stall speed, the control response is better than in a normal stall; and second, if the airplane is promptly rolled to a wings-level attitude the airplane is already out of the stall speed range.

Accelerated stall accidents also occur in the traffic pattern. They may happen on the first climbing turn, or when turning onto the final approach. In at least one instance that I know of, an accident resulted when an airplane on final was asked by the tower to make a 360-degree turn for spacing. Nothing wrong with that, but the pilot tried to hang on to his altitude by using back pressure instead of by adding enough power to compensate for the demands of the banked attitude.

The most valuable way to practice accelerated stalls and recoveries is to set them up the way they're likely to really happen. Begin by setting the power and trim for level flight but at approach speed. Now go smoothly into a 45-degree bank and attempt to hold your altitude with the elevator without adding any power. If you hold altitude, or even more surely, if you start to settle and then use the elevator without more power to try to stop the descent (*without letting the bank increase*—this is the hard part), you will stall.

The accelerated stall that results from abrupt use of the elevator is another one that really happens. It usually has no worse effect than to confuse the pilot, but sometimes it happens on final when an obstacle is suddenly spotted. Then the result of the stall is sharply increased settling and the airplane may strike an obstacle short of the runway. The basic cause of this accident, however, is the low approach, not the accelerated stall. Where it's confusing is in the landing flare, and here the chain of events may lead to a really bad landing.

When the pilot pulls back abruptly, suddenly realizing that the ground is only a few feet away, the airplane may assume a nose-up attitude so fast that the flight path can't change. The plane then continues down in an accelerated stall. The confusing thing is what happens next, after the descent is stopped by the ground, and the airplane begins to move horizontally along the ground. Now the wing is no longer at a stall angle of attack, and since the airspeed is still above the normal stall speed the airplane flies into the air again. We usually call this a bounce, but in fact it is a natural recovery from an accelerated stall. A "bounce" recovery is not hard to do, but for many people the natural response is to push the wheel forward to get back down—to point the nose where they want the airplane to go. This is not the right thing to do, because the airplane is sure to go down anyway and they don't want it to go down on the nose. But now they get a second touchdown on the nosewheel, and second and higher "bounce," and a worse situation.

I saw a classic case of this once, in which fortunately no

one was hurt. In this instance the last touchdown (the third) was so hard that the nosewheel blew out and the airplane ran off the runway and flipped over. The first touchdown and bounce hadn't been a bad one. If the pilot had just let well enough alone—held the wheel back or added a little power and maintained a slightly nose-up attitude—he would have landed nicely the second time. But from where I was, right next to the runway where he was landing, I saw the elevator moving just as it shouldn't. It was a perfect demonstration of what not to do and why. I wish I could make a movie of the picture I still carry in my mind's eye. It would make a great scene for a training film.

Solo Stall Practice

Most people don't like stall practice. The first time I did solo stalls I made up my mind to wait until I really had a stall before I recovered. Well, I did some stalls and they seemed a little violent, but I figured it was all in my head. Then I did one from a turn where the low wing stalled first, instead of the high one, as I was expecting. I really scared myself on that recovery. I used the wrong rudder first, responding mechanically to what I thought was going to happen instead of responding to what really did happen. I flew straight back to the airport, glad to be right side up and alive, and asked my instructor to come out with me and see if I was doing something wrong. It turned out that I was simply waiting too long to recover, and I was using the rudder too heavily and mechanically. The result was that I was getting incipient spins before the recoveries. He showed me how to keep the plane straight right through the stall with quick, alternating rudder.

And then we did spins. They were thrilling. I liked them. But it was many, many hours before I wanted to do any solo spins. And I still have absolutely no interest in doing spins in airplanes in which spins aren't recommended. There are surely hundreds, if not thousands of pilots who simply would not fly if they were required to do spins, especially solo spins, to get a license. It is all very well to wish that everyone had the kind of confidence that acrobatics can impart. But to limit the joy of solo flight to those people who do seems very wrong to me.

In principle I am in sympathy with those who advocate spins. That is, I don't believe that flying should be made to seem easy and safe just to get more pilots into the air. If I thought spins were necessary to make people safe pilots, I would demand them from my own students, and I would crusade for their inclusion in required maneuvers. But I don't think spin training is the answer to the problem. Instead, I think we need to look at the causes: overemphasis on flying by instrument indications, poor instruction, the pressure to get the license in minimum time, and the industry-wide overselling of aviation—the "Take a Drive in the Sky" approach. If the pilot is taught to be really comfortable, to center himself in the airplane and feel it as an extension of himself—and he can learn that if he takes enough time working at it—then he'll never need to know how to recover from a spin because he'll never spin. He'll recognize and break off the sequence of preconditions long before he gets anywhere near a spin entry.

The stall practice you'll get before your first solo will probably be very limited. You'll get more instruction later, and then you'll go out one day to do solo stalls. Those first stalls, all by yourself, will give you pause. You'll start a little higher than you've ever been before and you'll rehearse them in your mind pretty carefully before you finally make your clearing turns and take the plunge. This solo practice of stalls is a second milestone—not as exciting as the first solo, but worthy of some congratulations. There are a lot of licensed pilots around who've never done a solo stall, and never will. They are missing a lot of the joy of flying because they lack the confidence that must precede that joy.

After all this about stalls, let me say again, you can't stall unless you get too slow. And the airplane really will give you plenty of warning before you get too slow. A lot of practice on the sloppy edge of the stall as described above—coordinating power, rudder, etc.—will develop your ability to recognize marginal airspeed. This practice should be dual at first, and without reference to the airspeed indicator! Practice both level flight and climbs and descents with and without flaps and use full power too.

Are Spins for You?

Look closely now at how you feel about controlling the airplane. If a certain lack of confidence seems to arise from never having been in a spin, then you should certainly get spin instruction. Personally, I feel quite comfortable in a car without ever having had skid instruction, and I don't know how to do a maneuver called "double drift" either, but I do drive in such a way as not to need these things. I am confident that we have even more control of conditions in the air as pilots than we have on the ground as drivers, so I think there is no need for the ordinary everyday pilot to learn techniques which deal with situations at the edge of the realm of probability—provided that this ordinary everyday pilot learns to do a really good job of basic flying.

So think carefully about how you usually go about things. Do you tend to push into the unknown through a mixture of curiosity and pride? Are you going to be under social pressure, from family or friends, that could trap you into flying on the ragged edge of your knowledge and ability?

Maybe you do need spin instruction. If you feel that you do, I'd like to suggest that you get this work in an airplane in which you can also work on recoveries from an inverted attitude. Becoming unexpectedly inverted is far more unlikely than an inadvertent spin, but it could happen and knowing what to do will give you that much more confidence. If, on the other hand, you feel sure that your own nature and your attitude toward flying are such that you'll not need spin practice, I'm on your side. But do spend plenty of time practicing slow flight at minimum speed in various configurations. And once you're past the apprehension and can really observe what's going on, do practice stalls of all sorts to the point of boredom.

Taking Stock Before Takeoffs and Landings

By the time you've completed the block of lessons—probably about four hours—that comes before you begin takeoffs and landings, you have been introduced to all the normal things the airplane can do. Take stock of your position before you go on. At this point, before you start working in the pattern, you should not only be meeting private pilot standards as to altitude, heading, and speed control in the basic Here-to-There flying, but you should be able to analyze your own performance of all the maneuvers you've done so far. In other words, you should know what you're supposed to be doing, even if you can't yet do it very well. You should be able to look at your performance and figure out what's bad and what's good about it.

If you've been flying half-hour lessons, you are now way ahead of your counterpart who's been flying hour-long lessons. The first, obvious advantage you have is that you've started and shut down the airplane twice as many times as he has. You've seen the instructor's good landings (to use as models) twice as many times, you've interpreted the windsock, you've heard the radio exchanges twice as many times. You've also flown in twice as many of the infinitely varied combinations of conditions.

Beyond that, the short lessons give you more for the minutes you spend in the air because you're more alert and aware. Furthermore, if you get too much new material in one lesson, it can't be absorbed and begins to blur. Another point is that, because you'll wish your lessons were longer, you'll be thinking about what you've done, going over it in your mind, and you already know how valuable that can be. Finally, you'll be coming to each lesson feeling eager to get on with it.

Probably no one ever feels really ready for takeoffs and landings when they begin to do them, though eagerness may hide that unconscious uncertainty from some happy spirits. But if you feel very strongly that you need to review the airwork before you begin regular work in the pattern, let your instructor know. He doesn't want to let you have charge of the training process; that's his job and he's trained and qualified to do it. But he does need to know how you feel about what you're doing, and he should be willing to adjust his planned schedule and spend a little extra time going over the basic flying if you feel you need it.

7. *Takeoffs and Landings*

The hardest work your instructor has to do is helping you learn to make safe landings. There will be days when he will be praying for a blizzard, a hurricane, a flood—anything—just to get a rest from flying around and around the pattern. Each landing is new and challenging and important to you, the student, but there is a dreadful sameness to the pattern when you have flown it hundreds, or thousands, of times. There've been days when I've spent the time driving to the airport doing a trick in my head to fight off my own incipient boredom as an instructor. I think about how I felt myself when I was learning—the frustration when I couldn't see any improvement but had no idea what to do about it; the satisfaction when I solved a problem and broke through to the next level; the joy of my first solo. I can renew my enthusiasm for instruction by thinking about it from the student's point of view.

Helping the Instructor Help You

The worst feeling I know is what you feel when you aren't improving and you become aware that your instructor isn't succeeding in helping you, and you sense too that he is frustrated by his inability. The chances are the whole problem will be temporary and you'll soon be making good progress again. But if he seems to blame his own frustrated feelings on you, and if he lets you see that he's impatient, you will have a serious problem in progressing any further. You'll be too tense to really concentrate on awareness and perception. If you have that feeling, talk it over with him: "I have the feeling that you're angry with me for not doing better. Don't you know I'm doing my best?" Try to get a statement from him about what he is feeling. You'll both feel better if you talk it out. You may find that what you take to be impatience with you is only his feeling that he is failing in his job, that you are becoming discouraged and that it's his fault.

On the other hand, if he really is blaming you for not doing better, perhaps you can arrive at a better understanding if you discuss this frankly too. Maybe he thinks you should be flying more often and doesn't know about the reasons why you can't. Maybe he thinks you don't really care, and doesn't understand that you act lighthearted precisely because it means so much to you. If you both make a real effort to understand each other, and you still aren't happy, this is probably a sign that you're an ill-matched pair, and you should look for another instructor while he should look for another student.

Each instructor probably develops his own way of working on takeoffs and landings. I know I have, and your instructor isn't likely to be an exception. The last thing he needs or wants is a book-reading student telling him he's doing it all wrong. And probably his usual way will work beautifully with you; you'll make steady progress, and you'll solo quickly enough to satisfy both of you. But if you run into a period of slow or imperceptible progress and the frustration that can easily trend toward despair, I may be able to help, and your instructor shouldn't mind a tactful suggestion from you. I've worked with a lot of different students, in all levels from pre-solo to commercially licensed would-be flight instructors. I know that what works best with one person may not work at all with someone else. But I have found that there are some methods, ways of expressing things, and general approaches which seem to help almost everyone.

We all want the same things in a soloing student: a minimum safe level of proficiency; enough confidence to prevent incipient panic if the unexpected happens; the ability to analyze his own performance and to correct mistakes and gradually improve; sound judgment and self-discipline. I hope that some or all of the suggestions I make will be of value to you and your instructor in developing those skills and traits in you. But if he decides not to use a suggestion that sounds good to you, discuss it with him until you really understand why; he will have a good reason.

Getting the Most from Circuits and Bumps

There are a couple of things that are important if you are to get the most out of the time you're paying for. The first I've said before, but it's especially important when you come to takeoffs and landings, so I'll say it again. Fly half-hour lessons if you possibly can! If you are unfortunate enough to be learning at a busy, controlled field where you spend so much time getting into the air that a half-hour lesson just isn't worth the ground time involved, then you'll have to fly longer. But if the only reason for flying longer lessons is convenience, don't do it.

A second important point is another one that may present

a problem at some fields. Make your landings to a full stop. It is true that you may be able to squeeze ten or twelve landings into an hour of touch-and-go's and only eight into two half-hour, full-stop lessons, but the difference in numbers doesn't have the effect you might expect. You will make more progress in the two short lessons, and eight distinct, full-stop landings than you will in the one long, stressful hour. There are several reasons for this apparent paradox. Tension and fatigue are foremost. The touch-and-go allows a high level of tension to develop and tends to maintain it. You have no opportunity to relax and think about what you've just done and what you want to do next time. The hour also allows fatigue to grow until you will inevitably be practicing mistakes before the end of the hour.

There are other good reasons for not doing touch-and-go's in your takeoff and landing practice. First, knowing that you are going to be taking off immediately, some of your attention is diverted from the landing; but during the landing, the landing deserves your full attention. Second, in touch-and-go's the takeoff is usually initiated quite early in the landing roll, before the landing is really completed. This will tend to give you an unconscious feeling that a landing is completed when you touch down, and that can lead to trouble. "Loss of directional control" is frequently the cause of expensive accidents during landings. This loss of control usually doesn't occur at the moment of touchdown, but at some time later in the roll. You need to practice the whole thing and learn that you are still controlling the landing until you taxi off the runway. This will prevent you from relaxing your attention and your pressure on the controls too soon.

I'm afraid I used to have a habit that tended to encourage this very error in my students. I would say, "Beautiful!" when a student made a really good flare and touchdown; and I would say it just after the touchdown, long before the landing was really finished. The praise was heartfelt, and the student had a right to it, but saying it just that way and at that moment could foster the notion that the landing is complete at that point. I'm teaching myself to keep still now, until we've taxied off the runway and have stopped for the brief critique that follows each trek around the pattern.

Another reason for not doing touch-and-go's again has to do with habit. If you have a habit of "going" after "touching," you may inadvertently open the throttle after touchdown when you really intend to make a full-stop landing. This can cause you so much confusion that you may run the airplane off the runway and into a landing light while you're debating what to do. I saw that happen just that way to a solo student, and I felt that the error was not so much his as his instructor's.

A Basic Briefing on the Pattern

The briefing for takeoffs and landings is particularly important but is also especially difficult. Many pilots who fly very well can't put into words exactly what is happening during the approach and landing. I like to start by describing the takeoff, pattern, and landing in terms of the three basic performance elements you are controlling—speed, altitude, and heading. During the normal takeoff, climb speed and altitude are taken care of by using full power and proper trim, so heading is your main control problem. Airspeed must be monitored and corrected if necessary, but altitude can be ignored temporarily because the liftoff and climb performance you get will depend on the airspeed. Since you're already using full power you'll be controlling airspeed with the nose attitude.

Once you reach the pattern altitude, either just before or on the downwind leg, altitude moves ahead of airspeed in your priorities. You will establish some preselected power setting—usually cruise—and then control altitude with the nose attitude. This looks like a good place to briefly discuss the question of which control—elevator or throttle—should be used in connection with which performance element—airspeed or altitude. My answer is that it is most efficient to control whichever element is more important at the moment, usually the one that's being held steady with the elevator, following with power changes where necessary to control the other element. In straight and level flight you usually care more about altitude than about precise airspeed, so control the altitude with elevator pressure. In an approach or a climb the airspeed is fundamental to performance, so control that with the elevator.

As is so often true, the root of the controversy is a matter of what's appropriate. Power changes affect nose attitude very noticeably in light training-type airplanes, and so power changes generally require elevator use as well. For small corrections the elevator alone may work better. And if power is necessary, its effect is immediate. In heavy high-performance airplanes, however, power changes are not felt so quickly and a change in technique may be necessary.

Another source of this controversy is the attempt to separate airspeed and altitude in terms of control. It is not sound to think in an either-or way about these two elements of performance, because both are affected by both elevator and power. I once heard a nice story that illustrates the absurdity of thinking "either-or." An FAA inspector asked an applicant, "What controls airspeed?" He got what he felt was the wrong answer. He gave a long lecture on how the elevator controls airspeed and the throttle controls altitude, and then they went out to fly. After lining up for takeoff the applicant pushed forward on the wheel. "What are you doing, young man?" asked the puzzled inspector. "Trying to get up flying speed, sir," was the deadpan reply. A remark not calculated to put the inspector on his side through the rest of the test, but it does make a nice point about airspeed and altitude control.

Once you begin to set up the descent, airspeed becomes primary again and power and flaps are adjusted to control the descent while the elevator controls the airspeed. I don't want to get into the controversy about how to use flaps and power, whether a normal approach should be power-off or should be done with a gradual reduction in power, because there are so many factors to consider that it would take a whole chapter in itself. I'll just say that, all things being equal, I favor an approach at a constant airspeed and a gradual increase in flaps to steepen the descent, and I believe the entire final should be power-off.

Track (heading) remains important all the way around the pattern. Once the flare begins, altitude becomes primary again, but is so closely tied to airspeed control that they can hardly be separated. During the landing roll, power is off, the wheel is held back, and heading is the primary control problem again. And that's all there is to a circuit and bump.

This general explanation has been clear to most of my students, but the whole matter may become much clearer to you in the words of your own instructor. When I transitioned from the Cub to a Cessna 170 I had a lot of trouble with the flare and landing. My instructor kept saying, "Slow it down, slow it down," hour after hour. Finally one day my usual instructor was on a charter flight and I flew with another man who said instead, "Keep it off, keep it off." Suddenly I understood what I was trying to do, and I had no further trouble. So there is no one best way to explain these complex maneuvers so everybody will understand them.

Perception, Control, and Timing Equal Precision

Takeoffs and landings sometimes seem to get too much emphasis, and it is true that there is a lot more to flying safely than just making good takeoffs and landings. But the reality of the hard ground forces us to see these maneuvers as crucial. It isn't that it is any more desirable to be precise here; only that it is so much easier to see it when you aren't.

The level of precision in track and in altitude that's necessary in takeoffs and landings is what makes them so hard to do. The errors are so clear when the airplane is close to the ground that the pilot may tend to overreact or to react to the less important of the things he sees. Part of the difficulty seems to be that we don't easily see just what point the airplane is moving toward, especially if that point is not the same as the one the airplane's pointing at. That is, we don't judge accurately the actual path of the airplane.

The first thing you have to learn in takeoffs and landings is to see what the airplane is actually doing. A lot of elements in takeoffs and landings are dependent on perceptions that are difficult. The most troublesome are: 1) centering the airplane on the runway during takeoff and lining the airplane up with the runway during the climb (especially in a crosswind); 2) flying a final turn which will line up with the runway on the approach; 3) holding a straight final approach; 4) flaring at the right height; 5) holding the airplane just off the ground before the touchdown; and most difficult of all, 6) maintaining the angle of descent that will bring you smoothly to the place in the air where you want to start the flare for the landing.

Knowledge of proper control use and understanding of good timing won't do you any good, if your perceptions aren't accurate. Accurate perceptions seem to come very easily to some people and very hard to others. As I've said before, accurate perception depends very much on physical and emotional state. And I've also said that practicing mistakes is worse than no practice. So let me say again that it is really important that you cancel your flying if you have a cold, or a headache, or if you're upset about something. Of course, if your flying is a kind of release, a way to get away from it all, you may want to fly anyway. In that case, maybe you could return to the practice of less demanding maneuvers. Or, in extremity, you could ask your instructor to demonstrate a short cross-country. And remember, even if you're feeling well, fight that excess tension by shoulder shrugging and toe wiggling.

Lining Up on Takeoff

The first perception problem you'll probably have is lining up with the center line of the runway. Most people overcorrect for the fact that they're sitting on the left side of the cockpit. Because they know they should see the center line a little to the right of a point straight in front of them, they line the airplane up much too far to the left. Remember the cockpit is much more narrow than your car. During your initial training you'll find it easier, and you'll be more nearly right too, if you simply put *yourself* directly over the center line and accept the small error that results. When you're taxying, keep the nosewheel precisely on the center line. This will help you get used to the picture you see ahead when you're going straight. In time you'll be able to line up on the runway properly. And don't be discouraged because you're having a problem with something that seems as if it should be simple.

Even after you know where you want to be, you may have a little trouble with the takeoff roll the first few times. When you're trying to keep it straight, it may help you to know that the airplane will probably first try to turn to the left, then will run pretty straight, and then, just before liftoff, will try to go to the left again. To keep the takeoff roll straight you might try using only the right rudder. If you inadvertently use too much right rudder and the airplane starts to veer to the right a little, try just letting up on that rudder instead of actually using left rudder. Of course, that old bugaboo wheel steering may be a factor in any difficulty you're having here. If you are trying to steer with the wheel (unconsciously), you are probably slow to use the rudder, which is the only thing that will really work. Being slow with the rudder will mean, of course, that the correction, when you do make it, will have to be bigger. Work on correcting immediately, with the rudder, very gently, for very small errors in heading.

You probably have been making the takeoffs since the third or fourth lesson, perhaps starting by handling only the throttle or only the rudder, and by the time you begin working in the pattern you may already have solved the problems of the takeoff. It is in the approach and the landing that everyone has the most trouble. If you aren't sure you understand exactly what it is that you're trying to do, ask your instructor to explain and demonstrate again. Many students don't quite understand what it is they are supposed to see when we talk about glide angle or slope, or angle of descent. It may help if your instructor shows you both a steep descent and a shallow one.

Seeing the Glide Angle

I usually explain the glide angle in terms of imaginary flights of stairs. Try seeing it this way. After you turn onto final, imagine that suddenly three flights of stairs appear, extending before you down to the ground. One of them reaches the ground somewhere between you and the end of the runway

—short of the runway. One reaches the ground a little past the numbers, just where you want to flare for landing. One reaches the ground much farther down the runway. The slope of these imaginary stairways varies from steep through moderate to shallow. What you want to do is to keep the airplane on the moderate slope—the middle flight of stairs. So now, while you keep your airspeed constant, watch that slope. If the imaginary flight of stairs that runs from where you are to the place where you want to start to flare appears to be getting more shallow—flatter—your present flight path is too steep, and you need to correct by adding power. If the imaginary stairs are getting steeper, you must reduce power or add flaps to get down more steeply.

Most people seem to be able to see this more easily if the airspeed and the altimeter are covered. The instructor can cover both instruments using circles of contact paper with the backing only partly removed so he can see the dials but you can't. He will ask you what you think your airspeed and altitude are and will tell you when you're wrong. It generally takes less than an hour of this kind of practice to get to where you'll judge both airspeed and altitude very accurately just by sight and sound. Your control of speed will be smoother and will gradually become more and more nearly instinctive (reflexive), and erratic airspeed won't confuse the picture on your approach.

Once you have learned to perceive correctly, you have to learn how to use the controls to make the small corrections that your perceptions tell you are necessary. Because some of these are very small corrections, the use of controls is a little different from what you have been using for the fundamentals in the air. For example, once you're on final you'll correct the tiny errors in heading with your rudder, and the equally small errors in position with the aileron.

The final element, the thing that makes it work, is timing

your corrections properly to your perceptions. Even with accurate perceptions, and full understanding of how to use the controls, you won't be able to do what you want to do until you get the feel of the timing. And, like correct use of the controls, that can come only with practice.

What a Landing Really Is

The landing can seem like a big and unfathomable mystery. But once you get the whole business of the glide angle straightened out, it's really pretty simple. There is a common misconception that there is a kind of magic moment when you somehow know it is time to put the airplane on the ground and that putting the airplane on the ground is the landing. Actually a good landing is what happens when you move the controls as if you're trying to keep the airplane *off* the ground as long as you possibly can. One thing you can be sure of is that the airplane is going to get onto the ground sooner or later. There is no way you're going to be able to prevent that. What you can do is control the attitude the airplane is in when it touches down, the airspeed it has when it touches down, and (with skill and planning during the approach) the approximate place on the ground where it touches down.

The landing begins about fifteen or twenty feet off the ground (use a hangar, tree, or some other reference off to the side to judge that). This is where you want to start to change your flight path from downward to horizontal. If you can time your back pressure on the elevator control just right, you'll be within a foot or two of the ground when you are in a level or slightly nose-up attitude and are moving horizontally. This is the first half of the landing and a lot of people stop working at this point, and they usually get safe, but not full-stall, landings. A really good landing is made when you now continue through the second half of the job, slowly increasing the back pressure, raising the nose just fast enough to control the settling of the airplane, but not fast enough to lift it farther from the ground, until you run out of elevator and the wheels kiss the runway.

Perceptions in Landing

There are three common perceptual problems that affect the flare. First, you may not have any idea when you're at fifteen or twenty feet and so are making wild guesses about when to start the flare—using a different altitude each time—and are unable to find any consistency in what you see happening. Don't depend on developing an instinct for the right altitude. That will come in time, but it may take a very long time unless you are consistently using the right altitude. Look around, out to the side and ahead. Find a reference, such as a tree, house, or hangar. Observe what part of it you seem to be level with when your instructor tells you to begin your flare. Use that to tell you when to do it on your own. Don't worry about what will happen when you go to another field and the references you're familiar with aren't available. By that time you will have begun to develop a group of secondary, unconscious clues. They won't be as dependable as your familiar tree or flagpole or whatever, but they will help and with some conscious effort, picking a new reference that looks about the same, you should be able to do all right.

The second perceptual problem is knowing when you are within a foot or two of the ground during the second part of the landing. To help with this make use of taxi time. Learn to know just how the ground looks out to the side and ahead when you're taxying. It will look almost like that when you're a couple of feet off the ground. Incidentally, remember to do this consciously if you change to another airplane.

The third perceptual difficulty is the feeling that everything is happening much too fast. This effect is often at the root of what people call ground shyness—a tendency to pull the nose up too fast and so to "balloon," or fly upward slightly, instead of maintaining the float, or horizontal flight just off the ground. If you do this, it won't be because of some inherent fear of the ground, but simply because you are looking over the nose at the ground too close to the airplane. When you realize that it's time to flare, things appear to be moving very fast and you move fast too—much too fast. This is easily cured by looking all the way out to the end of the runway, and beyond to the horizon, then looking back to the runway threshold, out to the far end of the runway and to the horizon again, back to the threshold, out and back, etc. Keep your eyes moving, looking to the side too to judge height. But most of all keep looking way out ahead of you—everything will slow down remarkably, and you'll find you have time to perceive what you need to perceive and to respond as you should.

Timing in Landing

To most people it seems that it is the timing of the flare, from the point fifteen or twenty feet off the ground to the moment of ground contact, that is the tricky part of landing. But when you clear up the perceptual problems, you'll find that you actually have considerable leeway and time for adjustments. If you start the flare a little on the high side, you can adjust by pulling the wheel back a little more slowly than usual and can get a perfect flare and landing. Or if you wait until you're a little on the low side to begin leveling out, you'll have to raise the nose a little faster, but you can still get a perfect flare and landing. Of course, a large error in the timing of the start of the flare will require large corrections and the resulting flare can no longer be called perfect, but the touchdown may still be all you could desire. Once the flare is begun, small errors in the rate at which you raise the nose during the first half of the landing can be corrected by adjusting the timing of the second part of the landing.

I'd like to emphasize that a landing isn't finished until you turn off onto the taxiway. Think in terms of *flying* the airplane all the way down the runway, even though for most of the way the wheels will be in contact with the ground. This way of thinking helps to fight the tendency simply to put the airplane on the ground and is likely to help in maintaining control throughout the landing.

In Favor of Full-Stall Landings

There is a continuing controversy about whether or not to do full-stall landings. I favor full-stall landings, or at least a distinctly nose-high attitude when touching down and rolling out. If I fly with two pilots I've never seen before and Archibald makes a good full-stall landing, while Zachary makes a smooth landing but at several knots above stall speed, I have more confidence in Archibald's ability to perform well in all aspects of flight. Failing to make full-stall landings is simply a way of avoiding the consequences of not flying well. In many of today's light planes, which are designed to lighten the consequences of not flying well, the pilot who makes smooth-feeling but not full-stall landings may gradually make poorer and poorer landings without realizing it. Ultimately his failure to use the elevator properly will result in trouble.

Full-stall landings have a lot of practical advantages too, from longer wear for tires and brakes all the way to less likelihood of hydroplaning. Hydroplaning is a condition where your tires are skiing on the surface of the water on a wet runway, and your brakes will have no effect nor will you have normal friction to slow you down. This disconcerting condition, which has caused some accidents even in airline operations, depends on two things—speed and tire pressure. So keep your tire pressure up where it should be and your landing speed down where it should be.

If you reach a point where you believe you'll never make a bad landing again, you'll probably be wrong—even if you have thousands of hours. I've felt that certainty a couple of times, and I've unhappily proved myself wrong. Now it seems reasonable to *hope* that I'll never really crunch one on again, but I don't count on it. I work hard on each landing.

Unexpected Settling Close to the Ground

Certain conditions may cause you to start your flare too high. If you don't correct it, you will "drop in," running out of airspeed before you are down to the ground, then stalling and settling the last few feet. You may do it because you're overreacting to a tendency to start the flare too close to the ground. If that happens, it will probably be during your training and your instructor can help you take care of it. But later, after your timing is basically OK, you may raise the nose too soon—when you are too far off the ground—because you suddenly notice that the airplane is settling faster than it was, faster than you expected. This increase in descent rate may occur because you've only just now gotten all the power off, or because you just put more flaps down. Eventually you'll learn to have power and flaps all set up farther out on final. Until you do, this is one of those instances when the beginner has a harder job than the professional.

But this sudden faster settling within fifty feet of the ground may be caused, not by an error on your part, but by a change in wind speed. When you perceive the airplane settling, your natural reaction will be to pull the nose up to stop it. You won't see immediately that this is wrong and won't work, because pulling the nose up will actually stop the settling. But this effect is only temporary, because when the airspeed reaches its new low, you'll begin to settle faster still. Now you have a bigger problem than you had in the first place, because now you are both lower and slower than you were, and so you have less time and space to make the necessary correction. The perfect thing to do in each case, when the airplane suddenly starts to settle more steeply, will depend on your airspeed and altitude when it happens. For instance, if it happens when you are carrying an extra five knots for a strong wind, you may be able to safely ignore it, knowing you'll have enough airspeed to correct during the normal flare. But you can hardly go wrong if you add a little power and put the nose *down* a fraction. You'll feel as if this is only going to speed you right into the ground, but what it really does is assure that you will have the airspeed you need to flare with when you're ready to flare. If you find that hard to believe, try this: on final, with full flaps and your nose pointing down, add power and see how you sail right over what you were pointing at.

Fighting the "Six-Hour Slump"

After you spend some lessons doing nothing but takeoffs and landings you'll probably arrive at a point where you aren't doing even the simplest part of the pattern right—the well-known "six-hour slump," though it may occur at any time from four hours on up. I think I've found the cause for that and a way to clear it up very quickly. Here's how I figure it:

Flying is largely a matter of combining a lot of learned responses—essentially trained reflexes. A trained reflex is much faster and more efficient in terms of our nervous system than is motion, which requires the conscious direction of the brain. But when you begin flying, every motion you make in the airplane requires thought. At first you make a lot of wrong moves, or at least you're too slow with the right ones. Then one day your body seems to say, "OK, brain. I've got it now. Buzz off." Your brain cuts out of the system, but actually the reflexes aren't firmly set and mistakes show up that you weren't making when you were directing everything from your head. Your nerves and muscles have gone on automatic too soon. The cure is simply to go back, for another hour or two, to talking yourself through each bit of flying: "Left turn means left aileron with just enough left rudder, watch the nose, not too much bank, neutral on aileron and rudder, watch the nose, time to roll out . . ." You'll soon be ready to go back on automatic, at least for the fundamentals.

The first time you do the takeoff and the whole pattern and the landing it will seem as if there is so much to do in that brief time that you'll never be able to do it all without help. Yet at the end of even the first half-hour lesson of working on takeoffs and landings, it won't seem nearly so rushed. Somewhere in the second or third hour of takeoffs and landings you'll probably find yourself doing a funny thing: You'll start your glide much too early on the downwind. You'll do it because it feels like it's time. What happens seems to be this: Up until now it's been taking you the whole downwind leg just to get the straight and level squared away. As soon as you had that done, it was time to change everything again and begin the descent. But now, at least transitions are coming easier. The leveling out takes only a few seconds and you have it set up when you have barely begun to fly the downwind. You have a little time to relax and look around—but you never had any extra time before, and you feel you ought to be doing something, so you start your glide. When you make that error, be happy. It means you're getting ahead of things at least.

If you run into a period where you don't seem to be making any progress and neither you nor your instructor can figure out what to do about it, there are a couple of things that may help. First, think hard about the pattern, step by step. Then describe to yourself everything that happens from the time you taxi out onto the runway to the time you taxi off after the landing—what you see, feel, hear, think about, and do. If you can possibly do it, put the whole thing on tape and listen to it. Listening to yourself, see if you can pick up anything you left out when you were talking. Or maybe there's something on your tape that you don't seem to have time for in the airplane. If you make this tape and it sounds all right, but you don't feel confident that it's complete and correct, ask your instructor to listen to it and discuss it with you. He'll probably be glad to try anything that holds hope of getting you back on the track, and he may be able to pick up something from the tape that he hasn't noticed in the air —a misconception or confusion about what you're trying to do.

The other thing I've tried that may help is to fly a lesson or two where you and the instructor take turns doing the flying around the pattern. When he's doing the flying, you'll be telling him what to do—when to turn, when to start leveling out, when to begin his descent, use flaps, start the flare, etc. When you're flying, he'll be telling you as usual. It'll feel pretty strange to both of you when you are directing him, and he may have a lot of trouble actually waiting for you to tell him what to do when he knows you're making him wait too long (because waiting too long makes it so much harder to do what has to be done and stay within acceptable limits). Of course, he'll use his own judgment about how long to wait for you before he does whatever is necessary on his own. But this method may help both of you to see just where you're having trouble.

Depending on what part of the country you're in and what the season is, and depending on whether or not your instructor feels as I do about crosswinds and gusty winds for presolo takeoffs and landings, and depending on whether or not you are free to fly only on weekends, it may not be surprising if your training is interrupted for as much as several weeks during this phase. You'll be happier if it doesn't work out this way, but if it happens, don't worry too much about it. A lot of people come back from a break in the training flying better than ever before.

Some Valuable Extras

There are a few valuable demonstrations and experiences that too many pilots miss out on during their dual flying. One of these is a go-around from a point just above the runway with full flaps. If you've never done this, you may hesitate too long should an occasion arise when you actually need to do it. So try a few of these while the instructor is still with you. And most students, for some reason, don't go immediately into a normal full-power climb to pattern altitude if they initiate a go-around early in the approach—when they are still two or three hundred feet above the ground—so you should probably do that one a couple of times too.

Make at least a few approaches without any flaps, even if you are being taught (as I think you should be) to use flaps on every landing. If you have electric flaps, it is possible that the motor could fail or the fuse could blow, and you should know, just for the sake of your own confidence, that you can do it. It's not really all that hard. And if you possibly can—that is, if you can fly when there isn't too much traffic in the pattern—make a series of approaches completely power-off and get accustomed to how the airplane glides with the engine idling. Remember, you'll have to clear the engine during a power-off glide.

Another useful thing to practice is slow flight on the downwind leg, using approach speed, not minimum controllable speed. If you see another airplane close ahead of you in the pattern, it might be better to slow-fly while you're level on the downwind so you won't have to worry about overrunning him later. Being sure you can do that safely could make a lot of difference to you in a tight traffic situation. Finally, make a couple of takeoffs with the elevator trim set a little too far forward and a little too far aft. (In some airplanes this could be dangerous, but not in a light trainer.) Learn to compensate for this error, to apply the unusual but necessary pressures so smoothly that another person in the plane—the instructor, or a passenger—would never know anything was wrong. Of course, your instructor will discuss possible landing areas if the engine should fail at any point in the pattern, and will be sure that you know the proper procedures to follow, so I won't go into all that here.

Traffic and the Right-of-Way

Remember to practice the habit of watching for traffic. Look outside the pattern as well as ahead of you in the pattern. Make a special point of looking outside of you when you make your turns. The most important points are turning the corner onto the downwind leg, where a lot of people enter the pattern, and turning the corner onto final. There are still a lot of people who make straight-in approaches or who fly very wide patterns. The last chance you have to spot them, if they are outside and a possible hazard to you, is just before you turn onto final. You may seem to have the right of way in this case, but remember that the other fellow may have some kind of emergency and give way to him. There are some emergencies that wouldn't show—out of gas, loss of oil pressure, illness. So let him get on the ground and then find out if he had a good reason for cutting you out. You can size him up, and if he doesn't have a good excuse (and if he isn't too big), you can punch him in the mouth. But don't be like Mike O'Day of whom it was written:

> Here lies Mike O'Day.
> He died defending his right of way.
> His way was right and his will was strong,
> But he's just as dead as if he'd been wrong.

Becoming an Official Student

You can't solo until after you've passed a medical examination by an FAA designated M.D., so don't wait too long to make that appointment. After you get the medical certificate you'll begin to get a lot of mailings directed at pilots. This is a good time to consider joining a pilot's organization. I have always belonged to AOPA (Aircraft Owners and Pilots Association).* I don't love everything about the organization, and the magazine, *AOPA Pilot,* spends too much space on AOPA social and political affairs for my liking, but general aviation, especially the private pilot aspect of it, does need a lobby in Washington, and AOPA is the most effective one. I would count my membership fee well spent for that reason alone. Some of the people whom I most respect as pilots dislike AOPA and belong to NPA (National Pilots Association)** which also serves as a pilot's lobby. Do join one or the other.

*Aircraft Owners and Pilots Association, Washington, D.C. 20015
**National Pilots Association, 806 15th Street, N.W., Washington, D.C. 20005

The First Step in the Transition from Student to Pilot

When you begin to do takeoffs and landings the instructor will be telling you what to do as it comes time to do it. Then one day he won't do that anymore. Instead, he'll wait to see whether you know, by now, when to take each step for yourself. If you're thinking too much like a student and not enough like a pilot, you'll still be waiting for him to tell you what to do, and you won't take action even when you think it's time. He may let you know that he's going to start waiting for you to think for yourself, but then again he may not. So be ready, and start thinking in terms of taking responsibility yourself for the conduct of the flight. Do just what you would if you were alone. If you feel uncomfortably close to the plane ahead, and you'd go around if you were alone, go around. Your instructor needs to know how you would behave if he weren't there. Or, if you'd like to continue, knowing that he is there to get you out of trouble, because you want to see what would happen, OK. But let him know that this isn't the way you'd be doing it if you were alone.

Finally, the day comes. You solo. That is a feeling like no other, and I'm actually a little envious of you because you still have that marvelous experience ahead of you. I think I will never forget just how I felt driving home afterward—like the Empress of All the Russias incognito. I drove my battered old car home from the airport at its top speed of forty miles an hour, but in my heart I knew I wasn't really a forty-mile-an-hour-battered-old-car driver. Instead, I had become something indescribably glamorous and exciting. When a gang of kids passed me in their bright new Thunderbird, I thought, "If you only knew, you poor things. I look like a forty-mile-an-hour-battered-old-car driver, but that's only a disguise. I'm really an airplane pilot!"

There's something exceedingly heady about being the pilot-in-command—even for only a few minutes in a little Piper Cub. And as you go on and develop confidence and a habit of command, that confidence and self-reliance begin to extend into other parts of your life and they will make a difference in everything you do. Welcome to the club!

8. Get Ready . . .

For me, for many pilots, flying doesn't need to be useful to be justified. We want to be in the air, flying ourselves, and we don't really care whether it's practical or not. If we are going somewhere that we have to go anyway, that's a bonus, but most of the time when we fly from Here to There, we're only going There, wherever it is, because we want an excuse to fly. When we choose a vacation spot or a weekend retreat, it's because there's an airport nearby. We don't pick the resort first and then look for a convenient airport.

If You Need an Excuse for Cross-Country Flying

Of course, that's a slight exaggeration. In fact, a secondary but important joy of flying, if you're a family person, is the way it opens up possibilities for your whole family, especially if you have small children to be included in the fun. When my husband and I learned to fly, our children were seven, six, and four. An automobile trip was murder for all of us. We lived then about three hours (not counting traffic jams) from the New Jersey shore, but we had never been there. It was much too far to go for a one-day trip, and somehow too near for a vacation. But in the airplane we found our way there in less than an hour, with no traffic, no awful heat. We changed to beach clothes at the airport and walked the four short blocks to the ocean. There we stayed only as long as it was fun, before the sunburn and the crankiness began. We flew home refreshed and happy—all of us, kids too. They were bored by the flying part, but the flight was short and, besides, we let them chew gum, which wasn't usually allowed, so it was a treat. If kids are too small to chew gum, they usually go sound asleep as soon as they get in the air, and they don't wake up until they are back on the ground.

Now that you've soloed, you're probably beginning to imagine those trips you're going to make: family trips, or maybe with a date to have lunch in another city, or with friends for a day at the State Fair, or at one of the various attractions that have nearby airports. Three that come to mind on the east coast are Williamsburg and Luray Caverns, both in Virginia, and Callaway Gardens in Georgia. Or maybe you're looking forward to visiting that client who's five long hours away by car, but only a pleasant hour's flight in the plane. Whatever your dreams may be, you probably feel you've cleared the biggest hurdle to actually doing this kind of thing by making that first solo. And you have, but there is a lot more training before you'll be ready to set out on your first solo flight across the countryside to another airport.

Essential Equipment, Charts, and Other Things

The first essential for cross-country is a chart. AOPA (Aircraft Owners and Pilots Association) has a chart service that's terrific! This is definitely the best way to have at least one up-to-date chart of the area for which you take the service (a four-chart subscription—roughly two years). If you do a lot of flying, and if you draw lines on your chart as you should, then you'll need new charts even more often than every six months, which is approximately how often the new ones appear. Don't be cheap about it. If you can afford to fly, you can afford to use reasonably fresh charts. It's all too easy to find yourself way off course because you were looking at the wrong line, especially when you get a network of course lines drawn all over the chart. You can make a chart useful for more flights if you go over your course line with crayon or marking pen, using a different color for each flight, but when you can't make out the landmarks on the folds anymore, it's definitely time to buy a new one. And be forewarned, many airport operators run out of charts a month or two before a new one is due.

A subscription to the government's *Airman's Information Manual* is recommended. You may not want to continue to subscribe to all parts of it, depending on how much and what kind of flying you do, but many of the questions on the written test require that you use this manual and real familiarity with it will be a big help. There are a number of commercial subscription services which are designed to substitute for government flight information publications, or which present the basic information in compact form—easy to carry around and to use. I think the best of these for VFR use is *Sky Prints.* It's not very expensive as such things go and the monthly updating is easy. With this and Sectional and Terminal Area charts you have everything you really need for VFR flight. There are also some services which offer airport information beyond that available from the government. Jeppesen has both VFR and IFR service in a nice compact form, and there is at least one service available which includes photographs of a lot of the airports as well as diagrams. Try to look at as many of these as possible before you commit yourself to a subscription, because renewal is usually very much cheaper than initiating a new service.

You'll need a plotter and a computer. Look for the type of plotter which has a full-circle, rotating protractor with a grid on it. This can be aligned with a line of longitude or latitude or with an omni rose on the chart and you can read your true or magnetic course easily. Planning on the ground can be done easily with the fixed, half-circle protractor, but for getting a new course in the air the rotating grid is really a help. There's not a lot of difference among computers. Plastic ones, though cheaper, usually get melted or out of shape eventually. The big ones are easier to read, the little ones are easier to carry. Check the alignment and the ease of rotation before you buy. Some are too stiff and some are too loose and occasionally there's one that doesn't line up properly.

Some kind of clipboard will be necessary. You may want to use a plain ten-cent-store clipboard with or without a homemade leg strap of elastic or self-adhesive material. If you're going to do a lot of night flying, the light on a specially designed kneeboard may be really nice to have—a lot easier to handle than a flashlight, though you should have an ordinary flashlight too. If money is a consideration, and it is for most of us, start with the cheaper arrangement and see if you really need more. You could put it at the top of your Christmas list, if you find you really want it.

Over the years, gadget-minded pilots have designed vast numbers of little plastic objects to make flying easier. Most of them are more trouble in the cockpit than they are worth. For one thing they usually fall on the floor just when you want them. Maybe you have trouble figuring out the right heading for your downwind leg while you're entering the pattern. Believe me, it won't be easier when you're trying to fly with your head down, feeling around on the floor for the Handy-Dandy Downwind Finder.

For this particular problem there is a remedy in three steps that you should be able to do with no gadget other than your head; that is, if you still know how to add and subtract in this age of electronic pocket calculators. First, convert the runway into direction—i.e., runway 1 is 10 degrees, runway 10 is 100 degrees, etc. Second, if the direction is 180 or less, add 200; if it's 190 or more, subtract 200 (yes, I know you could get minus 10). Third, subtract or add 20, reversing the arithmetic process you used in the second step. For example, the tower gives you runway 13 for landing. First, you convert that to 130. Add 200 which gives you 330. Subtract 20, and you have the downwind heading—310. This is another thing you can practice on the ground.

The required instrument work, much of which takes place wearing a plastic hood that cuts out all visibility except the instruments, is torture for many students. One reason is that many hoods are very uncomfortable. The concentration during the instrument work is tough enough; you don't need to add any unnecessary discomfort. Your instructor may make use of the type of hood which is adjustable without elastic, and which is wide enough so it doesn't cut into your cheekbones. If he doesn't have one like this, it may be worth it to you to buy one of your own.

Before You Go Cross-Country

After the first time your instructor lets you fly around the pattern by yourself, you'll probably fly several more part-dual and part-solo sessions. And while you're doing that, your instructor may begin to cover some of the other things you need in addition to the obvious dual cross-country work he'll give you before he'll send you off by yourself. Depending on where and at what season you've been flying, you may already have an abundance of experience with crosswind takeoffs and landings, or with short and soft field takeoffs and landings. But if you've escaped these up to now, they are definitely in your near future. Most instructors don't send their students to airports where they need short or soft field techniques, but you'll begin to develop those skills now, "just in case." However, most of the practice of special takeoff and landing techniques will probably be deferred until after you've done some cross-country, so I'll return to those in a later chapter.

Practice Cross-Country Skills in the Pattern

While you're doing your "circuits and bumps," look at your home airport with a critical eye. Even though you know it well, you can practice an important cross-country skill right here: analysis of the approach (and the takeoff too) before you enter the pattern. Most pilots, licensed or not, don't take enough time looking over the field before landing at a strange airport. It is important to circle as many times as you need in order to really check out the area. Just stay well above pattern altitude.

Be concerned about obstacles, not only at the approach end, but also at the departure end and to the sides. At home you know where the wires are; suppose you were a stranger, could you see them? Can you see the poles they're strung on? Obstacles along the sides of the runway, a rise in the ground, a row of trees, buildings, can do some funny things to wind, even when the wind isn't very strong, and affect your landing. Here at home you may know about the peculiarities. Can you relate them to what you see when you look down? Even if you can't anticipate exactly what those funny things will be at a strange field, you can at least be ready for surprises.

Obstacles at the far end of the field have a bearing on where you set your go-around point, and you should always have a go-around point chosen before you begin the approach. Look at the area your instructor long ago told you was the acceptable touchdown area—usually the first quarter or third of the runway. See how long that section looks, and how much runway remains beyond it. The far end of that section is your go-around point. If you can see, when you're still at two or three hundred feet, that you'll be landing beyond it—overshooting it—you'll initiate the go-around right away. But if you think you'll be down OK, and then you find yourself floating, you'll go around when the go-around point goes by. This point needs to be chosen ahead of time, because once you are close to the ground and floating, you can't really tell how much runway is left. If you haven't picked some reference—a tree, or a hangar, or windsock, or something to tell you that Now is the Time—you're likely to wait too long. It's not a nice feeling to be praying that your brakes are better than you have any right to expect them to be.

Unless you happen to be learning at a really short field, you probably won't be sent to any really short fields when you go cross-country. But you may have trouble getting down properly at a strange field even of ordinary length—a lot of people do—so here's a suggestion. If you have to go around on the first approach, try using a modified short field approach on the second one. That is, use an airspeed a little lower than the normal approach speed and a little higher than the short field speed. Carry power to control the descent, setting up your flaps a little earlier than you usually do. This should give you more opportunity to judge whether your approach is too high, and will allow you to make enough correction (with the throttle, of course, controlling airspeed with nose attitude) to put you down on the runway just where you want to land.

Pros and Cons of Flying at Dusk

You may want to schedule some of this solo takeoff and landing practice late in the day when the air may be smooth and the wind light. But watch out for the period just at dusk. If you get at least some of your night flying experience before you go cross-country, you'll be a little ahead, but not if it comes on you unexpectedly and solo besides. Judging height, especially for the flare and landing, is most tricky in the twilight, even if the runway lights are already on. Your instructor probably won't let you out alone at that time; but if he does, or if you get back late from a cross-country, remember to be especially careful about looking around to judge your height. And another thing about practicing in that hour or so before dark: You may land and look around and feel sure that you have plenty of time for one more flight around the pattern before the sun sets. But remember to allow for the unexpected. Maybe the first time around someone will cut you out of the pattern; then you'll start back in and get cut out again; then the next approach is too high; the next time someone pulls out and takes off right in front of you. Anyway, by the time you get on the ground, there you are, in the dark, and you're not even sure where the light switches are.

Introductory Remarks on the Use of Radio and Instruments

If you haven't had some already, you'll now begin to work on using the radio for communication and navigation, and you'll be working under the hood on controlling the airplane by reference to instruments. Think of this practice as being like setting up guard rails in case you inadvertently get too close to the edge. Don't use these skills as people often do use guard rails, to lean on. Stay well back, even though they're there. That is, make your decisions and fly as if you don't have that extra skill. If you let yourself count on it, and figure it into your decisions, then the protection of having more skill than you expect to need is lost.

I don't want to say much about radio navigation now because of my bias in favor of pilotage, but here are a few things I hope you'll think about. Radio navigation is a useful tool—one that you can add to the methods of pilotage and dead reckoning in order to do a better job of flying a straight track over the ground. It was developed, not only to make cross-country easier and more efficient, but also to make it safer. Unhappily, the fact is that this capability may not offer you increased safety at all if you lean on it. What can happen is that, knowing you have radio, you may be led to fly into conditions of visibility in which you would never try to navigate by pilotage and dead reckoning alone. Think about this: When the visibility is poor, the ceilings are likely to be low too; and when you let down, to stay under those low ceilings, you may find you can't get any VOR (Very-high-frequency Omnidirectional Range) signal at all; or the one you do get may be unreliable or intermittent. If you're depending on ADF (Automatic Direction Finder), remember that obstacles and some kinds of bad weather may affect that too. So it's dangerous to think of radio navigation as a regular, everyday alternative to pilotage when pilotage gets too tough because of poor visibility.

Radio communication changes the whole aspect of life in the air. There was a time when we were really on our own up there. Now there is someone only as far away as the microphone. But here, too, low ceilings may make it impossible for you to reach anyone, or your communications may be too intermittent for you to get any real help. If you get in trouble in weather, start asking for help while you're still high enough to get it.

Using the Magnetic Compass When You're Lost

Many pilots get lost when the weather isn't a problem. Several years ago the FSS complained that over half of the lost pilots they gave headings to seemed to be unable to turn to the headings assigned or to hold those headings with enough accuracy to get anywhere, even when they were flying in VFR conditions. I gave that some thought and came up with a theory. I figured a lot of those pilots probably got lost in the first place because their directional gyros went bad or were mis-set. If the pilot didn't figure that out, he would still try to use the gyro for heading and would naturally have a lot of trouble. But I wondered if that could explain all the difficulty.

So I tried an experiment. I covered the directional gyro when I was flying with students and also when I flew on flight tests with applicants for the private certificate. Then I gave them a heading to turn to and hold—not under the hood, but VFR. All of them tried to turn to and hold the heading entirely by the magnetic compass with no attention to anything that they could see outside. And even though some tried to apply the compass correction rules, not one pilot was able to turn to a heading and hold it closer than about plus or minus 30 degrees. That's not good, and many did much worse. I decided it was necessary to teach specifically how to use the ground and the magnetic compass together to turn to desired headings and to hold those headings.

This is simply a variation on the 90 degree and 180 degree turns you practiced earlier and you can practice it by yourself. Fly straight and level and note your heading. Then assign yourself a new heading. Look at the compass and figure out how many degrees to the left or right you want to turn. Then look outside and estimate what you'll be heading for. Start the turn and use that outside reference to judge when to stop turning. When you're straight and level again and the compass has had a chance to settle down, check the new heading. If it isn't what you wanted—and you will probably be at least 5 degrees off and maybe much more—recalculate the degrees, reestimate it on the ground, and make another turn. Watch out for the common tendency to turn the wrong way. Hold the heading again by following what looks to you like a straight line over the ground, and check the compass every now and then to see if you need a small correction. Your instructor will show you how the magnetic compass swings back and forth, especially when your heading is within 45 degrees of 180 or 360. You'll see how important it is to hold the airplane as steady as possible to get a magnetic compass reading. And you should glance at the compass several times over a minute or so, averaging the readings if you get a different one each time, which is likely even in slightly bumpy air.

If I convinced you earlier that radio navigation capability can work like a trap, read on! Instrument competence, which the private pilot is required to have only because it should make him safer, can prove to be just as dangerous if it's misused. Knowing you have basic instrument competence may lead you to continue into deteriorating conditions of ceiling and visibility—conditions that you wouldn't try to fly in if the airplane didn't have any instruments in it. You will be leaning hard on a weak guard rail. All too many pilots break through and go tumbling into disaster.

New Responsibilities for the Solo Pilot

All along you should be developing confidence—confidence that you know enough, are skillful enough, to accept full responsibility for your flights. As always, that means practicing as you go along—judging wind conditions for yourself, and field conditions, and the condition of the aircraft too, even on dual flights. You must be able to take responsibility for everything from the pre-flight inspection you'll make to determine that the plane is airworthy and properly serviced to a final check to see that it's topped off and properly secured when you return to your home field. You should know what official documents are required to be in the airplane, and you should know where they're kept in the plane, and check to be sure they're really there. You should know what log book entries are required, and check the log books to see that they are OK. You need to know what kind of gas your airplane requires and what kind of oil is being used in the engine. And you must understand aircraft loading. Your instructor will supervise your departure. The real test comes not so much when you leave home with his blessing, as when you leave the strange field on the return flight. There you've parked, shut down, later pre-flighted again, planned your takeoff, started the plane, taxied, gone through the pre-takeoff check all by yourself with no supervision at all. You may not feel as confident as you thought you were. But if you've been taking responsibility for all that, you won't be quite so uncomfortable about it.

Malfunction Reports

To fulfill your responsibility as the pilot-in-command, you also must note and remember any flaws you notice in the airplane, or its equipment. Write down the missing screws, malfunctioning instruments, intermittent radios. And be sure that you write down exactly what you observed. Many a mechanic has removed one magneto and even taken it apart, only to learn that it was really the other which was bad. Learn how to make a good check of any malfunction you notice. Experiment with everything that might make a difference and report what you find. If you report engine roughness, for example, also report your altitude, power setting, and mixture setting. What did the carburetor heat do? What happened when you tried each magneto separately? Was the roughness more noticeable at certain power settings? Could it have anything to do with which gas tank you were using?

You may fly for hundreds of hours and never have anything to report, but that's not likely. If nothing else, you'll probably have at least minor difficulties with radios sooner or later. Radio difficulties are often not reported promptly in VFR training aircraft, and they may not be corrected immediately the way a mechanical problem would be, so several people have a chance to observe the malfunction, instead of only one. Anyway, if you have a problem, first make sure you really have a malfunction. On the omni for example, I've had people report malfunctions because they didn't understand the way the set worked. If you happen to tune in a VOR station when you are about 90 degrees off the course that happens to be set on the selector, you'll get an "Off" indication. You need to try turning the selector as well as tuning other stations before you report a malfunction.

If you can't hear anyone, there are a number of questions you should check and answer before you can make a full report. First, make sure the squelch control isn't turned all the way down. I turn it up until I get static, then down just below that point. Is the silence on both the speaker and the earphones? Is the silence on all frequencies—both communication and navigation? Did you check to be sure the mike button wasn't sticking, and did you try disconnecting the mike? Do you have evidence that the radio is working at all—that the problem isn't with the power source? (Navigation signal being received would constitute such evidence.)

If you call a facility and get no answer, consider the possibility that they haven't heard you. If nobody hears you, you sure won't get any answer. Listen long enough and to frequencies that are busy enough to be sure that the problem is really a receiver problem, rather than a transmitter problem or a mike problem. If you seem to have a transmitter problem—you can hear others, but no one will answer your calls—try a second mike if you have one, or a second radio. Check out all the frequencies you can reasonably try. Use the various FSS frequencies, Ground Control, Unicom, Air-to-Air, even emergency. If you definitely have a malfunction, check it out as fully as you can, and report it in writing when you get back to the airport. This effort on your part may save a lot of time when the radio repairman starts to work.

Planning for the Cross-Country

There are three more things you need a lot of information about for your cross-country flying: your route, your destination, and your weather. As to route and destination, at the very least your instructor will see that you have appropriate charts and that you check the facilities you're going to use in the *Airman's Information Manual.* But you may want to go further. If there is a local chart (on a larger scale) for any part of your route, you'll find more detailed landmark and terrain information on that than you have on the Sectional. And if your route comes within twenty-five miles of the edge of the chart, carry the adjacent one too.

There are several sources for runway diagrams, and you may want to look up your destination airport and see how it's laid out. Where are the taxiways, if any, and which side of the field is the tower on? One source includes photographs of a number of fields, and several indicate wires, trees, buildings, etc. If nothing else is available, you can ask your instructor to let you use his instrument approach plate for your destination (if the airport has an instrument approach). The approach plate has a diagram that shows runways, taxiways, height of significant obstacles, position of wind indicator and tower and other buildings. You might even want to take a road map. No kidding! I took one on all my student cross-countries.

Weather Information

Of course, the official source for weather information is the Flight Service Station or the U.S. Weather Bureau. But don't ignore television. Some of the morning and evening news shows provide maps which can give you a very good general picture. And on Thursday and Friday nights public TV in many places offers an excellent broadcast covering aviation weather for the weekend all over the country. The very best source for weather in your area will depend on what facilities are nearby and how well staffed they are. In general, the local FSS is the best. One of its principal reasons for being is to provide the pilot with appropriate weather information. Its people are specially trained to understand what weather means to the pilot and in most cases are able to do a better job of briefing than the U.S. Weather Bureau.

Know precisely what you want before you make the call

on the phone. Some stations want you to give the information on your flight in a one, two, three, four order, after which they'll give you what they think you need. These facilities are understaffed, and the person you're speaking to is under a lot of pressure to cover more work than one person can possibly do. He may have pilots calling on several different frequencies, several different telephones, and people arriving in person at his counter as well, so be sympathetic. But remember too that when the guy behind the counter says, "It looks all right to me. I don't think you'll have any trouble with thunderstorms," he's not going to be up there with you if he's wrong. It will be your life, not his, that may depend upon the judgments you make. Make sure you get all the information that will make it possible for you to evaluate the more up-to-date and important information you'll get later from your own eyes.

You need first a general picture—highs and lows, fronts, and general movement and weather associated with the major features. Of course, you don't need all that for a fifty-mile flight to be made within an hour or two of the time you call. But if you should be lucky enough to get someone who isn't too busy, you could get it all just for practice in putting it together. Then you want a forecast for your departure point and your destination, and perhaps for several stations along the route and off to both sides if it's a long flight. Finally you want the most recent reported weather. Some briefers object to giving you both forecast and present weather, but persist. You can sometimes get a good idea of how reliable the forecast is by comparing the present weather with what it was forecast to be earlier. You want winds aloft, of course, but ask for surface winds too. Winds aloft are often very inaccurate, and surface winds may be a help to you in judging what's really going on, both as to probable winds aloft and as to weather movement. Ask if there are any pilot reports that are pertinent to your flight.

Take the first opportunity you have to visit an FSS in person. Learn the identifiers for the reporting weather stations in your area, so you can find them easily and read the sequences and the terminal forecasts. Practice taking down the weather in standard symbols. When you get weather for a flight over the phone, *write it down,* using standard sequence and symbols. Then later you can compare it with current reports and get a good idea of the trend. If you don't write it down, you may not notice later if there is a clear indication of much faster-than-forecast deterioration, or of much slower-than-forecast improvement. And if you get weather reports while you're in flight, try to write those down too. If you can reach one of the automatic weather broadcasts on the ADF while you're on the ground at your home field, you can get some practice listening to that. I won't suggest an FSS frequency; they seem to be changed pretty often. Check the AIM *(Airman's Information Manual)* or ask your local station which are the best ones to monitor in flight.

Above all, relate what was forecast and what is being reported to *what you actually see.* There are places and seasons when visibility reports of two or three miles in haze, reported in the morning from the field on the city's outskirts, bear no relation at all to what you have at your field fifteen or twenty miles away, where you can see thirty miles at pattern altitude. Then there are other situations when a report of "better than six" for visibility seems like a cruel joke when you get up to 2500 feet and can't see anything in the golden soup that's passing for air up there. The briefer knows only what comes over the teletype, or what is available on his own instruments, and that information is limited. Over a period of three or four years, provided you observe the weather carefully and compare what actually happens with what is forecast to happen, you should accumulate enough experience to be able to put together what you see with the official forecast and come up with more accurate local forecasts of flying weather than the briefer can make—unless the briefer is also a pilot and has been in the area as long as you have.

When it comes to en-route weather in unfamiliar areas, you'll have to depend more on the FSS and the Weather Bureau, because you won't know the significance of specific local conditions; but as you fly, you come into possession of information the weather people don't have—that is, you know how it looks right now. Always try to combine what you actually see with the educated guesses that you've been given by the professionals and form your own opinion about what is happening and what's likely to happen next. Of course, you'll be safer if you tend to be slightly pessimistic. If you and the briefer disagree about the outlook, take the worse of the two predictions and make your Go–No-Go decision on that basis. And remember, you can help everyone by making pilot reports yourself, if you encounter conditions that are different from what was expected.

Keeping a Flight Log

You haven't been keeping in-flight records up to now, nor made written plans. Now you must begin. There seem to be almost as many forms for planning sheets and flight logs as there are pilots, or at least as many as there are flight schools. Look over as many as you can find and see what you like. For instance, you will need space on the flight log for in-flight remarks. I find it's easier to write on the top of the page than on the bottom when I'm using a lapboard or kneeboard, but the blank space on most flight log forms is at the bottom. If you design and run your own sheets off on a Xerox you can rearrange to suit yourself.

I personally like to use the chart itself to keep my log,

making a separate planning sheet when I feel I need one—usually for a long flight, requiring more than one chart, in an unfamiliar area. I write my estimates and actual-time-over on the chart. And I mark off measured segments ahead of time. It reduces the cockpit clutter, and I can follow the chart for navigation more easily if it isn't tangled up with a lot of sheets of paper. But for the kind of record-keeping that your instructor will probably want from you—the kind he needs so he can give you a good constructive debriefing after your cross-countries—you'll need a separate flight log.

If all this sounds like a lot of work, that's because it is. But the rewards of flying a well-planned, well-executed cross-country flight are enormous. You feel like a very special person—and you are!

9. Get Set . . .

Yes, yes, I imagine I hear you saying. No doubt all this is important. But what about actually navigating? All right, let's start with some general discussion of navigation before we finally talk about getting into the air and doing it.

A Couple of Specific Physical Problems

There is an interesting problem in navigation for some people that may be almost insurmountable. It has to do with what seems to be one particular kind of intelligence or cognitive ability—dealing with spatial relationships. Although we don't know much yet about the factors that determine intelligence, we do know there is a physical basis in some people for a deficit in ability to comprehend maps, to visualize space and form. If you are one of those who suffer from this problem, you will have to expect that cross-country will be difficult. The fact that your instructor will probably be impatient if he doesn't understand the nature of your problem, will make it all the harder. And it may be that he won't understand. The chances are that you are at least as intelligent as he is, in general. And since many of those who have this problem have exceptional verbal ability, which is how we often judge the intelligence of others, he may feel that you can't possibly be so dumb about cross-country unless you just aren't trying. It may help if you explain that yours is a special problem.

I believe you will be able to learn to fly safe cross-country flights, but you won't do it in quite the same way as most of us. You will have to translate what seems self-evident and commonsensical to most of us into specific verbal rules that you will have to memorize, and follow in an automatic sort of way. And you should be even more careful than the rest of us about flying cross-country in less than optimum conditions or at a time when you aren't up to par physically or emotionally. You should be extra sure your flights are never long enough to cause real fatigue, and avoid circumstances that could put pressure on you in the air. Of course, there is also a normal variation in ability in this aspect of flying. Navigation is easy for only a few lucky souls who seem to have that special bump of direction that others lack.

A pair of much more common and obvious physical deficits give a lot of students difficulty—nearsightedness and farsightedness. If you need glasses for all vision and are accustomed to wearing them, you probably won't notice any hardship. But your field of vision may be slightly reduced and you'll need to move your head more in order to take in all that you should see. It is those who use glasses only for near vision, or only for far vision, who experience the most trouble as students. If you have to put on or take off glasses every time you look from the chart to the ground or vice versa, pilotage can be a real drag. If you use bifocals and are accustomed to them, this may not be too bad. If you use reading glasses, or glasses only for distant vision, you'll find life in the cockpit a lot easier when you hang your glasses around your neck.

Learning to Think in Terms of the Compass

Those are things you can't change about yourself; you can only compensate for them. But there are other things that may make the practical business of learning to navigate more difficult, and these are things you can help yourself with. Some people have never been taught the basic compass directions. There are even some of you I've discovered—perfectly intelligent, no special cognitive difficulties—who don't know that the sun rises in the east or that most maps are printed with north at the top. You just aren't accustomed to thinking in terms of direction, beyond the simple and self-relative concepts of right and left. Of course, it will be harder for you than for those who, like me, learned about directions when they were so young that they can't even remember when it wasn't natural to think in those terms.

You can teach yourself the skill of orienting yourself in relation to compass directions by constantly practicing it. In your home, in your office, when driving or waiting at a red light, figure out north, south, east, west. Picture yourself on a map facing whatever direction you really are facing. Also try "eyeballing" direction. That is, draw some lines on an old chart. Estimate the direction of each one. Then check it with your plotter. Keep doing that until you can consistently guesstimate within 10 degrees of the correct direction.

Navigation, Especially Pilotage, Is Hard Work

Even for people who easily *learn* how to navigate, doing it is still a lot of work. But if you're fortunate, the work will be so interesting and the challenge of solving the cross-country puzzle so stimulating that the effort feels like play. This effect—work seeming like play—isn't peculiar to flying; a lot of effort is often put into what we call recreation. Look at skiing, tennis, or golf. But in skiing, tennis, or golf you can go on for years just fooling around. In flying you don't have a choice; a lot of effort is *required.* This is especially true as you are acquiring the knowledge and understanding and experience you're going to need in order to navigate accurately and efficiently.

Navigation methods have changed a lot through the years. The altitudes, speeds, and distances have changed. Our expectations with regard to reliability and safety have changed. Especially in the last twenty years or so there have been enormous changes. When I began to fly in 1957 there were plenty of VOR stations, at least in the populous northeast, but very few pilots were yet learning to fly in airplanes equipped with radios. Demonstration of the ability to use radio for communication and navigation wasn't included on the private flight test until 1958. Every instructor in those days knew how to navigate by pilotage, though there were some who did not know how to use a radio, or who knew how to use only the low-frequency range or the ADF. Radio was a luxury—the sets cost money and so did the power sources. A training plane—Cub or Aeronca—wasn't likely even to have an electrical system, let alone radio. So the student (and his instructor) knew that when he was up there alone, he'd really be alone—no way to ask anyone for help until after he got himself back on the ground. You can imagine that a lot of time and effort went into learning to navigate by pilotage.

Today it's very different. The pilot doesn't expect to be really alone in the air, even on a solo flight. Every trainer is radio-equipped. Every instructor knows how to use radio—in fact, every instructor has to have an instrument rating. It sounds great. Surely well-trained instructors and well-equipped airplanes should be making cross-country flying better and safer. Alas, they could, but they generally don't. Pilots increasingly depend on electronics to get them out of trouble. With radio they are able to avoid the disasters that would otherwise follow on the heels of careless planning and sloppy flying.

Yes, Virginia, Pilotage Is Possible

Suppose your instructor feels pretty uncertain about pilotage. He's not sure he can do it—not even confident that it's possible to do. He's never seen anyone navigate by pilotage in an unfamiliar area. Neither had his instructor before him, or his instructor before that. You can hardly expect him to spend a lot of time trying to teach you what he thinks may be impossible. Even if he knows pilotage is possible, he's likely to believe that it's not practical in modern airplanes, at modern speeds, especially in those with low wings. So your instructor will make sure you get a good background and plenty of practice in using VOR and he'll want you to use visual checkpoints, but he probably won't teach you the various little things that make pilotage really practical—even valuable and interesting—mostly because he doesn't know them himself.

Actually, you, as a student and as a low-time pilot, are the one who is most likely to need skill in using pilotage. It's a matter of economics. Radio gear is expensive. The rental airplane you fly as a student and VFR licensed pilot, and even the airplane you may buy—either alone or as a member of a small club—will probably have one or at most two radios, one power source, one speaker, one mike, and very likely no earphones. The equipment will be fair to good, but it won't be the best you can buy, because if it were it would cost more than the airplane. The best is reserved for the business and airline fleets where IFR flights are the rule and equipment breakdown is costly. When the whole reason for the existence of an aircraft is to get paying customers from A to B, the reliability has to be very high. So not only do these fleets have the best, they also have backups that we, in the private sector of general aviation, can't afford—two gyro-compasses, alternate vacuum sources, alternate power sources for the many radios, extra mikes and headsets. Unlikely as breakdowns and malfunctions are—and all approved instruments and radios are remarkably dependable—it is the inexperienced pilot who is most likely to run into a problem. How much of a problem it is will depend on how well-developed his skill in pilotage is.

You're in luck if your instructor happens to like pilotage and is good at it. If he doesn't and isn't, you may find some resistance if you want to spend much time on it. But you can still work with him as you learn it, if you approach him tactfully. Be careful about the way you go about asking to do some cross-country using only pilotage, or only pilotage and dead reckoning. Describe it as a personal idiosyncrasy and beg his indulgence. You'll get a lot more cooperation than you will if you say, "Listen, you ought to teach me this."

The Theory of Cross-Country

Before you go any further in this chapter, review all the terms that relate to cross-country: track, course, heading, LOP, fix, radial, bearing. It is essential that you understand clearly the difference between track and course and heading, between radial and bearing. Maybe it's all jargon, but it's jargon you've got to be able to use easily. These words all have special and important meanings in the world of aviation.

Think of a cross-country flight as a puzzle: The solution to this puzzle—what you're looking for, using the methods of navigation—is a perfect heading. This perfect heading, if you could find it, would keep you on a straight line between your departure at Here and your destination at There. It would correct for the effects of wind and for the peculiarities of your compass. It is an ideal, and the challenge of navigation is to come as close to it as possible. We have three separate navigation methods we can use when we're trying to find that heading.

➳ *DEAD RECKONING.* The first method is dead reckoning. This is pure theory—that's why I call it "first." You do it on the ground before the flight. Of course, you can plug in new information on wind and speed as you fly and make new calculations. You start by finding the true course; applying wind factors and airspeed; getting a true heading; converting that to magnetic and that to the compass heading; figuring estimated elapsed times. You'll get all that in your ground school work, and there aren't any special tricks to it. The practice is a little more complicated when you apply new knowledge acquired in flight—ground speed, distance off course—but the actual method can be practiced in a comfortable room with pencil and paper until you've developed a high level of skill.

➳ *PILOTAGE.* That new knowledge acquired in flight depends partly on the second method: the practice of pilotage. Pilotage is basic and practical. Pilotage is the first way anybody goes anywhere—empirical, no theory at all. Pilotage is a matter of flying by landmarks alone. Even a compass and a clock have no place in pure pilotage. Pilotage is how we walk across a parking lot—we see where we want to go and we head that way. It's how we follow directions to get to a friend's house—"Go right at the gas station, turn left at the second street, and it'll be the first house on the right." My boss used to like to say that the way to get to Florida was to go to the first ocean and turn right. That's pilotage.

➞ *COMBINING PILOTAGE AND DEAD RECKONING.* But when we talk about the solution to our puzzle in terms of heading rather than just in terms of "getting there," we're demanding something that pilotage alone can't produce. We must have dead reckoning to combine with it—we must use at least the compass. So we're going to combine these two methods. We'll start out with the heading our theoretical planning gave us. Then, after we've held the heading a little while, we'll compare our real position over the ground with the theoretical position we find on the chart under the course line. Now we can see what error we have in our theoretical solution—the solution we got by deduction (dead reckoning). Then we make changes in our heading (and maybe in our time estimates too). After holding the new heading a little while, we check again to see whether it is any closer to the solution to the puzzle.

The combination of these two methods works well when there are plenty of clear and distinctive landmarks indicated on the chart. It works well at fairly low altitudes, below clouds, and at relatively low speeds. But there are thousands of square miles that aren't distinguished by any landmarks, or at least not by any that the government wants to put on the chart. And a long flight is often much more comfortable if you get up above 5000 feet where the air is likely to be smoother, even if you are now above scattered or broken clouds. So you might fly for twenty or thirty or forty miles, and have no clear references to tell you that the wind has changed and that the puzzle solution you found earlier now has to be worked out again.

➳ *RADIO NAVIGATION.* This is the third method. In effect, radio navigation provides a continuous checkpoint. You have a constant reference to tell you whether your heading is the one that solves the puzzle. Actually, it can't be quite that precise, but it is usually more nearly continuous than pilotage can be. There are exceptions, where landmarks do provide a constant reference for correction. You can use a coastline, highway, river, or mountain ridge, if you happen to be going along one; and the grid of section lines in the central United States can be used for flight in any direction if you develop a good eye for angles. But most of the time the use of the radio will enable you to do a better job of precision cross-country flying.

➳ *SOME SPECIAL ADVANTAGES OF PILOTAGE.* I am still in favor of pilotage whenever possible. For one thing, if you want to use it at all, you need to practice it or you may find the skill has gotten rusty just when you most need it. There is always the possibility of losing the radio. Second, even if you have the radio, usually you won't be able to fly a straight line from Here to There using the VOR, because you will seldom find a transmitter that lies precisely on your course. Third, and most important, when you fly pilotage your attention is outside where traffic, obstacles, and bad weather could be lurking to astound and dismay you.

Just think of all those people, flying VOR, all heading toward the very same spot in the sky—right over the station. It's an alarming picture. If all those pilots are doing a good job of radio navigation, they are paying a lot of attention to the indications in the instrument panel. Even if they were all trying to maintain a good inside-outside scan, and I know there are some who aren't, I'd rather not be in that airspace just over the VOR station. I don't mind it much if I can be there on an IFR flight plan and actually in the clouds. Then I can feel pretty sure there's no one else in there with me. But in VFR conditions I just can't feel comfortable.

Look at the Line from Here to There

Cross-country of any kind begins with planning. The first step in the planning is drawing the line across the chart from Here to There. Take a general look at the area. What kind of terrain is it? What's the highest elevation, and the highest obstacle along the way? Look for brackets—terrain features that will be very distinctive such as rivers, mountain ridges, and large cities. They will tell you that you've wandered off course either to left or right, or that you've flown beyond your destination. If you can find good ones, brackets provide a nice sense of security.

Once you have the broad picture, begin to look at your course in more detail. Study the symbols indicating landmarks you'll be looking for when you fly that course. A common mistake is looking only for landmarks that lie directly on the course line. Your planning will be more valuable if you think of yourself as flying in the middle of a corridor ten miles wide. You should be able to spot any good landmarks that lie anywhere in that corridor, because you won't be flying cross-country unless you have at least five miles visibility. Don't ignore major landmarks that lie farther off—cities, lakes, mountains. They may be useful too.

Using Linear Checkpoints

Now you want to choose points for time checks, so you can correct your estimated elapsed times. And you'd like to be able to use the same points to judge whether you're off course, and if so, how far off. These should lie under or very close to your course line. The ideal thing to use is the intersection of your course line with some linear feature that crosses it, preferably at an angle close to 90 degrees. That could be a road, a railroad, a waterway, a coastline. Of course, you won't always be able to find such points, at least not enough to serve as all the necessary checkpoints along your route. But wherever you have one, use it.

The advantage of the linear feature crossing your course compared to an isolated point is that you can use it if you happen to drift to one side or the other of your intended track. If you're off course, you could miss seeing that isolated small town, or tower, or lake. But if you've chosen a road that crosses your course, you're going to cross the road—even though at the wrong spot. You'll be able to get a time check for future reference, and you should be able to use the road or whatever the checkpoint is to help you return to the course. Thus, with the linear checkpoint you can know not only that you aren't where you intended to be, but also where you are. Then you can figure out the necessary correction. Notice, though, what happens if the crossing angle isn't close to 90 degrees. Even a small error off course to the side will change the distance enough to put your original time estimate way off. You'll have to remeasure the distance to get a fair estimate of ground speed.

The Best Checkpoint Isn't a Point

The term "checkpoint" can lead you into a kind of thinking that doesn't make for sound pilotage. The things you're looking for are almost never isolated points. Instead, a checkpoint is usually at least two features which are related to each other in space, and more often it is a whole group of features which lie in distinctive relationships to one another. For instance, suppose you are using a creek. You need to get more specific by relating it to something else—the creek at the point where it crosses the highway, or the creek at a point one mile west of the railroad bridge. If you're using a town, the town can be distinguished from other towns by related features like highways, railroads, creeks, lakes, towers, racetracks, drive-in movies, etc. Careful attention to the whole pattern of features will keep you from misidentifying a "checkpoint." Make a practice of penciling a circle around the groups of landmarks you'll be looking for, instead of just drawing a line through the point you'll use for your time check. It will remind you that you're looking for the complex of related features, not just for one point.

Seeing the Stuff on the Ground

This is where that cognitive disability affecting spatial relations will make it tough. Most of us can look from the chart to the ground and *see* if they match, but if you have this cognitive problem you may have to express the complex of points verbally and sequentially and match things step by step. For example, you will say to yourself, the town has a tower beyond it on the right (look out and check it); there is a road that goes out from the town and to the right of the tower (look and check); there is a creek that crosses the road beyond the tower, etc. This will be harder and also slower than what the rest of us are doing but it can be done. If, added to this problem, is the peculiar difficulty some people have in using the terms "left" and "right," normal pilotage may be impossible for you. If someone out there is accomplishing it in spite of these difficulties, I'd like to hear how he manages to do it.

It is one thing to study the chart, understand the symbols, and know what you'll be looking for. It is quite another to see them and identify them from the air. This is where most people need help, and where constant practice pays. I once had a colleague who complained that he never could see any big yellow cities when he looked for them on the ground. Of course, we all know cities aren't yellow, but do you know that water often isn't blue?

There is some disagreement about whether natural or man-made landmarks are more useful. Well, it depends on where you're doing your flying. If your area is pretty open and not much interfered with by man, then the man-made features will stand out most. But if you do your flying over miles of city, suburb, parking lots, superhighways, and shopping centers, natural landmarks may be more distinctive. This is another case where it's a matter of what's appropriate. With some practice you'll be able to find most of the things on the chart most of the time. But things often don't look as you might expect them to.

➻ *THE YELLOW TOWN.* A town which is shown by a yellow shape on the chart is supposed to be big enough to see as that shape at night. Usually the day shape corresponds with the night shape. The shapes on the charts, however, don't keep up with development. It may be that the whole set of things looks right—everything but the shape of the town which doesn't match the chart. Look more carefully at the part of the town that doesn't match. If it is just too new to be on the chart, it will probably be made up of residential developments and the trees in the yards and along the streets of these new developments usually will give them away because they'll be small.

➻ *THE RAILROAD.* They used to say IFR meant "I follow railroads," or they called the railroad the Iron Beam. Railroads are still good landmarks. They usually run in pretty straight lines, and you can check the angle the railroad makes with your track as you cross it and see if it looks right. You can usually spot the railroad right-of-way before you see the tracks because it shows up across open land as a line of bushes and underbrush and across wooded land as a cleared line. Once you develop an eye for railroad tracks you'll find them as useful as roads for day flying.

➻ *HIGHWAYS.* Roads are obvious and valuable. The only trouble is that there are so many more on the ground than there are on the chart, at least in most parts of the country. And around major cities there are so many it is almost impossible to make use of them with any confidence. It will help to remember again that they are not isolated but are part of a complex of features. You can distinguish between turnpikes and other major highways by what is associated with them. Turnpikes will have very limited access, with toll booths visible. Interstate highways and freeways also have limited access, but these generally don't have any facilities along them, while turnpikes do have occasional service areas. Secondary roads either come to a dead end or go over or under these highways. Ordinary dual highways have gas stations, diners, discount houses, and motels along them, and secondary roads run right into them.

Roads are usually easy to see; it's just deciding which road you're looking at that may be hard. I once flew a helicopter to pick up a person in a very exclusive community in Long Island. I had a well-marked road map but I couldn't find the place—an estate where I was supposed to land. My problem was that the roads I saw on the ground didn't seem to match the map at all. I finally landed in an open space where I saw some construction workers. One of them was willing to go up with me to show me where the house should be. It turned out that the roads I could see weren't the roads on the map. The public roads in that old and elegant town were completely hidden by enormous old trees, and it was only stretches of the winding macadam private roads on the estates that I had been able to see. When my construction worker friend got me to the right place the whole family was out on the enormous lawn waving sheets and bright umbrellas to attract my attention.

➳ *MISCELLANEOUS LANDMARKS.* Keep in mind that everything that's on the chart is there because someone thought it would be helpful in air navigation. The tiny black squares represent various landmarks of particular value. In the middle of Death Valley a single tiny building deserves a black square. A more common use of these square black marks is to designate institutions—schools or hospitals. The feature that usually helps you spot these is the water tank and/or stack that's likely to be associated with any institution big enough to make it as a charted landmark. Drive-in movies may be shown as a black square or, more helpfully, they may appear in a small symbol that shows the orientation of the parking area. This lets you know whether or not the movie screen will be a good landmark for you. If you are looking directly toward the screen, you'll be able to see it easily and from surprisingly far away. If you're looking at it sideways, or from the back of the screen, it'll be pretty easy to miss altogether.

➳ *AIRPORTS.* How about airports themselves as landmarks? Some are excellent, used in conjunction with other features. Others are very hard to spot. But even when you have plenty of other landmarks, and aren't planning to land, practice finding the small grass fields you fly over. Even the smallest grass field usually has buildings next to it with access to a road. And look for a building that could be a hangar. Look for a windsock. Above all, look for parked airplanes. It is possible to develop air eyes, just as hunters develop what they call woods eyes. When you get your air eyes, you'll be able to spot grass airports as fast as most people can spot Kennedy International, and who knows, there could come a day when you'll be glad of that talent.

➟ *THE QUEEN OF LANDMARKS—WATER.* Finally, my nomination for the most consistently useful terrain feature! Water! There are several reasons for my enthusiasm. First, waterways and coastlines and lakes are usually charted very accurately, and they are not much subject to alteration. Second, almost all cities of any size are built next to sizable bodies of water or on waterways, so a habit of using such features in navigation will often be very helpful in orienting yourself around unfamiliar metropolitan areas and busy controlled fields. Third, water often can be seen at some distance in haze when other things, even cities, can't yet be distinguished. Finally, in more sparsely populated areas charted streams often are not only the best but also the only features available, so it makes sense to know how to spot them.

What I meant when I said that water often isn't blue, wasn't so much that it may be silver or gray, or even gold or red at sunset, or white with snow in winter, but rather that it's quite possible you won't be able to see the water at all. Still, you'll know it's there because where water runs or stands there will usually be distinctive vegetation. You'll see that the trees are a different kind and a different color, or that the dusty plain is broken by a clump of trees, or by a straggling, meandering line of green. Even where the ground is intensively cultivated, usually you will find the running water easily because farmers know that streams run more steadily if they are lined by trees; so what you'll see is a narrow band of woods in a crooked line that matches the blue line on the chart.

Practice finding the streams that the chart shows in your area. Look for that pattern of vegetation that matches approximately the line the chart shows. Someday the ability to find and follow these lines with your eye may help you find a small grass field just when you need it most.

10. Go!

Now you're going to get into the airplane and really do it—fly from Here to There! Get above the traffic, the common herd. You won't be there alone at first, but it's still a taste of what you've been aiming at. Or so the theory goes—cross-country is the payoff. It's true that a lot of us choose a destination only to give us an excuse for flying, but cross-country flying is not, therefore, to be thought of as incidental. This kind of flying is an art. To practice any art you must first master the crafts it utilizes. To practice the art of cross-country you must first master the basic craft of straight and level flight. If your craftsmanship in this area is sloppy, the attempt to practice the art will be hardly worth the effort. But let's suppose you can handle the demands of the craft. Now you must learn to put it all together and go somewhere. Theory is one thing; applying the theory when you're actually in the air is another.

Judging Distance over the Ground

One basic difficulty a lot of people have is judging distance. First, to use pilotage effectively you need to be able to look at the chart and know, easily and almost without thinking, approximately what the distances are between points on the chart. There are two pretty simple ways to develop your eye for this. First, use the marked control zones as a rough ruler. The typical circular portion of a control zone is ten miles in diameter. Second, use the minutes of latitude which are indicated along the lines of longitude. Each minute is equal to one nautical mile. Be careful if you're using a chart with a scale different than the one you're most accustomed to. At first glance it may confuse you, but if you use the diameter of control zones and minutes of latitude as distance scales, you can soon reorient yourself.

Once you know in numbers what the space on the chart represents, you have to be able to relate those numbers to what you see when you look out of the airplane. This takes some practice, so whenever you're in a precise position you can positively identify, and you see an object on the ground at a distance that you can measure on the chart, take a good look and file away the mental picture. But notice your altitude too; distances look different at different altitudes. The higher you are, the closer together things below you will look.

There's a way to estimate distance over the ground that helps a lot of people. Use the runway at your home field as an imaginary yardstick. Picture that runway in your mind's eye, and estimate how many times it would fit into the space between you and the landmark whose distance you want to judge. Then you can work out a formula, using the known length of your runway, to get a rough estimate of the distance. For example, if it's about half a mile long, simply divide the number of times it will fit into the space across the ground by two to get the mileage between you and the landmark. If your home runway is about two-thirds of a mile long, divide the number of times it will fit by three and then multiply by two to get the mileage. And so on. At least this will give you somewhere to start when you want an estimate.

Drift

Once you've learned to look through the windshield straight in front of you instead of over toward the center, you'll expect that it will be easy for you to tell where you're going. Strange as it may seem, that isn't always the case. Drift is the problem. Drift is a kind of motion that man seems to have no natural talent for distinguishing. Unlike animals that swim or fly, we don't encounter it in our everyday lives. It is hard for us to see and understand and cope with it. For those of us in the air it's caused by wind whose direction is from the side relative to our heading. But, because we are not attached to any fixed object—such as the earth itself—we are *not pushed* by it. Rather we are moving *within* the mass of air which is felt as wind by groundlings. We actually float *with* it as we fly *through* it. This motion can be hard to see, especially at cruising altitude, even when there is enough drift to affect seriously the track of the airplane. If it should happen that there is a noticeable drift at some time during your early dual flying, it will help if your instructor takes some extra time to point out to you how the drift is affecting your track. And it will be much easier to see drift if you are close to the ground, so it may be worthwhile getting down to approximately pattern altitude to look at it.

Here is a method that may help you to see the drift. Look ahead and line up two objects in front of you. Now hold a heading that keeps the more distant object ahead of you.

Give it a couple of minutes. If the nearer object seems to move to one side, let's say to the left, or if you have to turn to the left to keep the more distant point straight ahead, you are drifting to the right.

Now you want to figure out how to correct for the drift you see. We'll assume the drift is to the right. Look toward the horizon, not toward what's straight out in front of you, but a little to the right. You're looking for the same effect that you used to spot the drift in the first place. That is, the nearer objects move to the left of the more distant ones. You'll find that there's a direction you can look—maybe 5 or 10 degrees to the right of straight ahead—where the objects stay lined up. That track—a line toward those objects—is the track you're "making good," and your drift angle is the angle between that track and the point straight ahead. To correct for the drift, to move toward the objects that appear to be straight in front of you, you'll have to turn left by about that same angle. Then the objects you want to pass directly over will be visible off to the right at an angle equal to your drift angle. This isn't as complicated as it sounds, so study it and try it in the air.

It can be made more confusing than it needs to be if your instructor talks to you about "feeling" drift. The fact is that you can't feel drift, and neither can he, but both of you may think that you can. What we can see happening we seem to need to feel as well, and even when we can't feel anything, we imagine that we can. We've all had the experience of sitting in a parked car and grabbing for the brake when the guy parked next to us started to back out. We "felt" we were rolling. We think we *feel* motion because of what we *see.* Drift is a motion that can *only* be perceived visually. If a fog rolled in under you, or if you were flying over a featureless sea, there would be no way for you to know that you were drifting. So don't try to feel it; do try to see it.

Getting Off Course and Correcting It

There is really only one way to get off course; you have to fly the wrong heading. But there are reasons and reasons for flying the wrong heading. First, you may have unexpected drift you don't notice right away, or no drift when you planned for and are correcting for some. Second, you may carefully hold a heading based on a plan that incorporates some error, such as failure to include the necessary compass correction. Or you may hold the correct heading but use a gyro that isn't set correctly. Finally, you may simply not fly the heading you mean to fly with enough accuracy to hold you on a straight line.

Once you get off course, there are two ways to make corrections that will take you where you want to go. One, and the most efficient, is to make a correction that will enable you to fly the new course directly from your position to your destination. A better way, if you're a beginner, is to make a correction which will take you back to your original course line and then another correction to stay on it. But it is important to avoid making extreme changes in heading. Let the angle of your drift or error determine how much correction you make. You can estimate that angle roughly on the basis that one mile off course after six miles of flight equals a drift angle of ten degrees. (One mile off course after ten miles would be six degrees.) If you correct back using twice the off-course angle, you'll get back on the course line in approximately the same time and distance it took you to get off it. Then you can reduce your correction angle to the same as your estimated drift angle and you should just about hold the course. Notice, though, that this will work only if you can hold accurate headings. You see how the art depends on the craft.

At Your Destination

In a very real sense, your arrival at your destination is the most critical part of the flight. You are more tired now than you were at takeoff, and you have the pressure of having to control the airplane while you look things over. Every student, in fact almost every pilot, is in too much of a hurry at this stage of the flight. Make it easy on yourself and take your time. Think in terms of "cheating"—doing as much as you can before you get there—studying a diagram, figuring out where you'll be in relation to that diagram as you approach, and planning how you'll circle and enter the pattern. And then double-check everything while you circle the field well above the pattern and out of the way.

While you are circling the field there are several things to look for besides traffic and wind indications. Check for markers indicating right hand traffic, even though you should know about that already if you've used the *Airman's Information Manual* or some equivalent source. Check the area for obstacles at each end of the runway as well as along the sides. If your runway has a hollow or a body of water off the approach end, plan a slightly steep approach to allow for the possibility of settling when you get close to the threshold. If you see a road across the end of the runway, the chances are very good that there will be wires. Find landmarks around the field—roads, houses, etc.—to help you fly a good square-cornered pattern.

➻ *BE WARY OF RADIO INFORMATION.* Listen on the Unicom or the tower frequency well ahead of time, or to ATIS (Automatic Terminal Information Service) if there is one, and try to get a clear picture of how the traffic is moving. That won't always be possible. Don't be dismayed if you hear the Unicom say, "Landing at pilot's discretion." This probably means that the person on the radio can't see the field and doesn't know with any certainty what is happening out there. It's better for him to say "pilot's discretion" than to give you a landing direction which may be wrong but which you take as Gospel. When given a landing direction, too many pilots relax, don't check wind indicators or other traffic, and come in the wrong way just because that was what the radio said. At a controlled field be wary about making assumptions before you receive your own clearance. You may happen to hear an instruction given to another pilot who, for some reason, is not using the same runway you'll be instructed to land on.

➳ *THE WIND AND THE RUNWAY.* The first thing you're looking for is everything that can help you decide which way to land, and there's a lot! In fact, there's so much that it can get confusing. To begin with, you know what the wind was expected to be. If the wind is given as "300 at 10" that means it's blowing *from* 300 degrees, and you would expect to land on runway 30 (spoken as "three-zero"). Your heading on landing would be 300 degrees, so you'd be heading directly into the wind. In the very early days of aviation, runways were numbered according to different rules at different airports. Runway 1 at one field might be the first runway built. At another it might be the longest. But soon it was agreed all over the world to number runways according to magnetic direction. Landing on runway 1 you are on a magnetic heading of 10 degrees; landing on runway 10, you are on a heading of 100 degrees, etc.

As you fly, you'll be looking around and watching for indications of wind direction: flags (on post offices and schools); clothes on lines; bending grain in fields; leaves on trees (the side of the tree that looks silvery is the windward side); ripples in ponds and lakes (the water along the windward shoreline will be calm); and smoke (but be careful—it can be misleading).

➟ *AIRPORT WIND INDICATORS.* At the airport itself you may have as many as three different indications and those three may conflict with one another. First, just about every airport has a windsock. It is usually mounted on a pole somewhere near the runway, or up on one end of a hangar building. Like smoke, it may be misleading unless you check it from several angles, because of the effect of foreshortening. And if it is close to trees or buildings it may give erroneous indications for some wind conditions even if you are reading it right.

Then there may be a mechanical wind indicator—a T and/or a tetrahedron. Some fields leave these swinging free, others lock them either automatically or manually, when the wind is light or is blowing directly across the runway. It is possible that the wind may have shifted or increased since the indicator was locked, so the indicator may be indicating the wrong runway.

Both the sock and the wind indicators give a lot of people trouble in interpreting them correctly. The large end of the sock—the end attached to the pole—is the windward end. Some people like the image of landing as if you were flying out of the large end of the sock. If the airport uses a wind T, interpretation is easier. It looks like a small airplane heading directly into the wind, just the way you want to be when you land. But remember the locked-T problem mentioned above. The tetrahedron may be confusing because here the small end, the pointed end, swings into the wind. It points toward where the wind is coming from. Think of it as an arrowhead pointing the way you should land. And remember this, too, may be locked. Go over all this and be sure you have it straight, because it's important to the safety of your landing.

➟ *OTHER TRAFFIC.* The thing that should be, and normally is, the best indicator of landing direction is other traffic. But it's not safe to make assumptions about that either. Be on the lookout for the heavier airplane, maybe a light twin, whose pilot may elect to use a crosswind runway because of its length, under wind conditions in which you'd be better off on the shorter runway, heading into the wind. Or maybe the airplane you see is being used specifically to give advanced crosswind dual to a commercial or instructor applicant, and the crosswind component is too much for you in your airplane. Notice in connection with these indications that it isn't enough to look for traffic in the pattern for the runway you select, even if that's the right runway. If more than one runway is available—and there's always the other direction on the one you're using—look around carefully for possible traffic on another runway.

At some fields there is a hard-surfaced runway but no paved taxiway, and traffic may taxi on the runway. Here you might look down, see an airplane rolling on the runway and assume that it has just landed, when actually the plane you see is taxying back after landing the opposite way. Incidentally, it is important if you land at such a field to roll out to the end of the runway before turning around. If you stop and do a 180 before you reach the end, you may find yourself nose to nose with another plane whose pilot assumed you would roll to the end. And, if you are the incoming pilot, play it safe the other way, and don't assume the plane ahead of you will roll out to the end.

Finally, there is the possibility that the pilot of the plane you're about to follow made a mistake when he selected the landing runway. Students and low-time pilots tend to assume that everyone else in the air is more experienced and more right, but this may not be true. Even experienced pilots sometimes make assumptions based on earlier wind conditions, or they misread the indicators. The results may range from merely embarrassing through expensive and humbling to completely disastrous. And the results may be more far-reaching than the effects on the pilot and passengers.

One summer evening at Flying W the county Democrats were holding their annual picnic on the field. We were trying hard to put our best foot forward, hoping to convince some of the local politicians that aviation is safe, fun, profitable, and here to stay. There wasn't much traffic, even though it was a lovely evening. We'd taken some of the picnickers for airplane rides, and now they were enjoying their barbecue. After a sultry day with southerly winds, a breeze from the northwest had just begun to cool things off a little.

And then it happened. A fellow flew in with his four-place airplane full of friends and misread the indicator. If he had just ignored it and landed on the long runway in either direction he would have had plenty of room. Unfortunately, he didn't do that. The indicator was lined up with the short runway, so he used the short runway—the wrong way. That meant he was approaching over wires as well as landing slightly downhill with the wind behind him. We learned later that his approach speed was extra high too—he thought that was safer. Anyway there was no way his brakes could stop him in time, though he pulled on the handle so hard he bent it, and the airplane went over on its back at the end of the runway. Happily, there were no serious injuries, but the plane was severely damaged and there were other hidden costs. Whether he or any of his friends were scared right out of aviation I don't know. But I do know that the summing up of such accidents makes everybody's insurance rates higher. And, of course, the Democrats saw the whole thing as evidence that flying is dangerous, and our public relations campaign had a severe setback.

The moral is: Take everything into account, and then make your own decision about what is best for you and your airplane. The usual rule is to conform to active traffic, but as you can see there are exceptions. If you do elect not to conform to the flow of traffic already established in the pattern, you must allow the right of way to that other traffic and make your approach so as not to conflict with it. As an instructor flying in a pattern with other instructors and students I have often faced the problem of trying to get the pattern turned around when the wind changes. Sometimes it's taken quite a while! There are advantages to flying at controlled fields, or with the radio on all the time. Finally, remember that even when you're given the runway in use—whether at a controlled or an uncontrolled field—the final responsibility for the safety of your flight is always your own.

It's gratifying to make a really good approach and landing at a completely unfamiliar field. With careful preparation on the ground before your flight, and deliberate planning in the air over the field, you can have that satisfaction almost every time.

11. *With Everything You've Got*

The complete job of cross-country includes the use of the radio. Any tool that can make cross-country flying safer and more efficient must be welcomed, even when there is some risk that it may lure some unwary pilots into a dead end of cross-country incompetence.

When I argue for more attention to pilotage, I feel like a person standing on the beach saying to the tide, "Don't come in!" A former student recently asked me what I thought about advice he had been given (by his instrument instructor) that he scrap his Sectional charts and carry only Radio Facility charts instead. Could he have been teasing me? Alas, I'm afraid not. If he was, he certainly got what he was expecting —an extended oration on the joys and values of pilotage.

Understanding VOR

As I've watched pilots being trained with more and more emphasis on radio and instrument flying every year, I've expected to see some loss of pilotage skills. And I haven't been surprised. But I didn't expect to find that a lot of the pilots who have no idea how to fly pilotage also do a very sloppy job of radio navigation. There are even instrument pilots who do not use the omni to solve the cross-country puzzle—to find that perfect heading more efficiently. Instead, they use it to make the careful holding of a heading unnecessary. Omni lets you wander back and forth between the point where the needle swings far to the left and the point where the needle swings far to the right without any fear of becoming hopelessly lost—unless or until the radio breaks down! I once flew with an instrument-rated pilot on a VFR day who alternated happily between a heading of 30 degrees and a heading of 90 degrees when flying a course of 60 degrees. I wish I could believe that that was an isolated case, but I'm afraid there are a lot of pilots who do no better.

Among the earlier omni sets was one that had a little window in the face of the dial in which you could read "to" or "from." If the set was off, or if you were too far from the station to get a reliable signal, the word in the window was "off." I heard an old-time instructor say with disgust one day that they ought to print "lost" in place of "off," because most pilots were lost when their sets were off. Too true!

The Implications of Angular Error

Understanding the omni and how it works needs a lot of diagrams and pictures, not to mention a lot of practice. I'm not going to start out and go all the way through the process of using VOR for navigation, but there is one point that I particularly want to get across; you can follow the diagrams to see how it works. The point I want to make is that the error which an omni set can show you is angular error. Angular error is the error in terms of degrees off course. Degrees must be measured from a point somewhere, so you actually have an indication of degrees off course as measured from the VOR station you've tuned in. There is no direct way that your omni can give you error in terms of distance, and you'll have to do very impractical things, such as turning 90 degrees off course and timing the change in indication, in order to find out what your error is in terms of miles off course.

Let's look at some of the implications here. First, notice that all four airplanes in the first diagram will show the same 10 degree error. A, B, and C will have a left needle indication and will indicate "to." Airplane D will have the needle to the right and a "from" indication. It is important to understand that if each of these airplanes could suddenly turn into a helicopter and start hovering on the spot while spinning around like a radar antenna, these indications would remain precisely the same. Almost everyone supposes at first that the indications on the set, that is, "to" or "from" and the left or right needle, are in terms of the nose of the airplane. To correct this misconception, it may help to think about it this way. The omni set has no way of "knowing" what direction you're heading. When you set the course selector it is as if you're telling the set that that's the way you're heading—that that's the line you want to be on—so all the indications will be correct *only* if that's approximately the direction you really are headed.

Now let's suppose that you're in airplane C and you start out flying on the course line. You find yourself over in position C_2 in diagram 2 because you drift with the wind. You now change your heading, by some lucky chance, just enough to stay on the line that will take you straight to the station, but you neglect to "tell" the omni set that you've changed your plan a little; that is, you let it "think" you still want to fly the original course of 0 degrees, because you don't change your omni bearing selector. The needle then will continue to show exactly the same error even though you'll be flying a correction that's going to put you back on course and directly over the station. Making this kind of correction can happen only by chance. It doesn't solve the cross-country puzzle; you haven't found the perfect heading to keep you on your original course.

The way to solve that puzzle is to use bracket headings—a farthest-left limit and a farthest-right limit, and as you experiment with headings, work the two limits closer and closer to each other. These limit headings, of course, will have to do two things. First, they must return you to your course if you drift to either side. And, second, it will be inevitable that if a heading brings you back on course, it will also take you through the course to the opposite side if held too long. With practice you should be able to get the spread between the two headings down to 5 degrees.

Let's see how you'd go about doing that. Take the case we've illustrated. Your original heading was 0 degrees and that allowed you to drift to the right, so 0 degrees immediately becomes your first farthest-right bracket heading. You now know you're looking for a correction for a wind from the left and you know that 0 degrees is a heading that will return you to your course (and through it) if you should happen to get out in left field. Now, what will you try for the farthest-left bracket on this first attempt? A fair start is a 30 degree cut, though that must depend in part on how far you are from the station and how many degrees you allow yourself to drift off course before beginning your correction. Let's assume that you try a heading of 330 degrees and it corrects

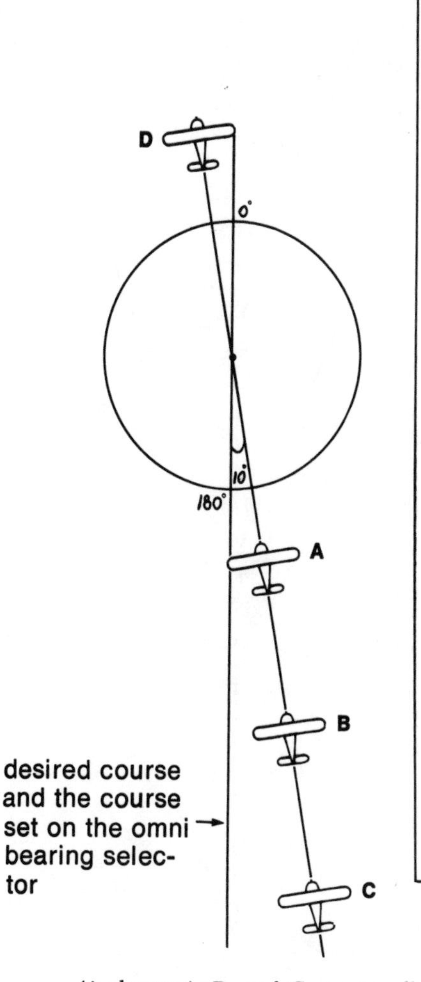

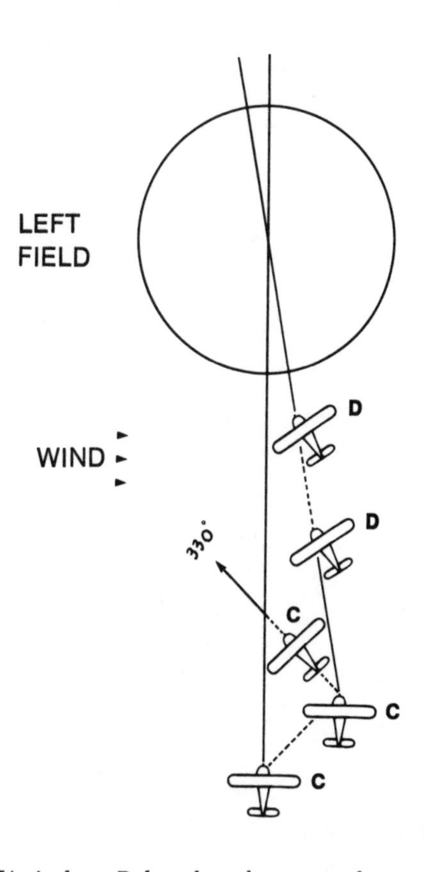

If airplane D has the selector set the same as C, his needle will indicate a constant error all the way to the station.

Airplanes A, B, and C are equally far off course in terms of angle. The omni needle will hold steady if C flies to position A and will not show that a correction toward the course is being made until the needle centers directly over the station. If the plane continues to D, the needle will again hold steady but to the left instead of the right.

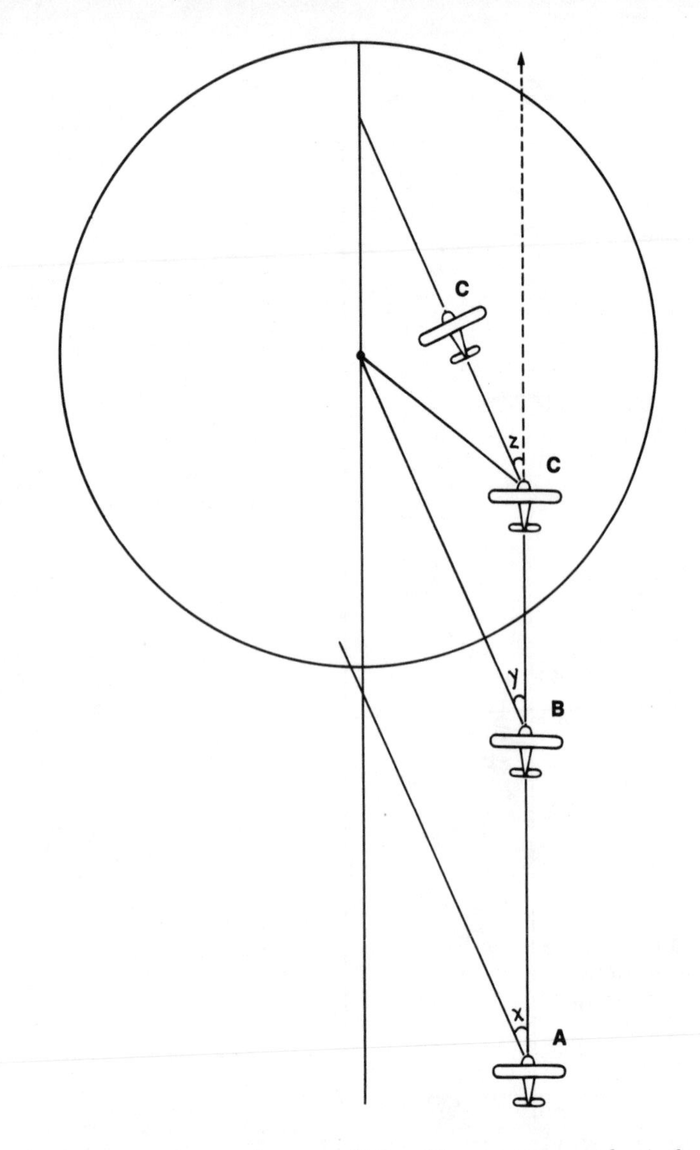

Angles x, y, and z are the same. Using this correction angle airplane A is back on course before reaching the station, B crosses the station, but C will reach the course beyond the station. During the correction leg A's needle will slowly center, B's needle will show constant error, and C's needle will first show increasing error. Pilots in C's position usually overcorrect.

you OK. Back on the course line with the needle centered, you now cut that 30 degree spread in half and fly a heading of 345 degrees.

Suppose you now drift to the right again. Then you know that your perfect heading, your puzzle solution, must lie somewhere between 330 and 345 degrees, so you mentally change your farthest-right bracket from 0 to 345 degrees, and fly 330 degrees, as you did before, to correct to the left. When the needle centers this time, you split the difference again, trying 338 degrees. Suppose that this time, using a 338 degree heading, you don't drift to the right with the wind, but instead move to the left. OK, that means that you've learned that the wind doesn't require that much correction, and you know that your perfect heading lies between 338 and 345 degrees. Of course, you won't use 330 degrees anymore, but will now change the left limit to 338 degrees. You still have a spread of seven degrees, 338 to 345. This will be more than you'll want when you get really close to the station, so after you turn right and fly 345 degrees until the needle centers, you split the difference once more, trying 341 degrees. You've now reduced your spread to 3 degrees if this heading takes you to the right, or 4 degrees if it takes you to the left.

The degree of craftsmanship you've developed in flying straight is very important here. And you've got to make your corrections quickly, when the angular error is small, only 1 or 2 degrees, or this method won't work well. If you wonder about that, look at the third diagram. Study this to see that the small corrections won't work if you wait too long to apply them. If you are in airplane C_1 and correct to 340 degrees, you are making a good cut at your course and will get back on it well before you pass the station. If you wait until the angle between your course and a line from you to the station is bigger, as at C_2, and then you try 340 degrees, your needle can't indicate, at C_3 for example, that you're getting closer to your course. This is because the angle between what you have set on the selector and the line you're on relative to the station is getting bigger, even though you are closer in terms of distance. The needle can't start to show that you're correcting until after you pass the station.

What usually happens when a pilot gets in this spot is that he increases his corrections to such an extent that he shoots right through his course line and out the other side before he can turn. The keys to flying omni efficiently are: 1) to work on reducing the bracket spread—to work, that is, with a plan in mind, with the goal of finding the smallest possible spread within which your perfect heading must lie, and 2) to make your corrections at the first clear indication of error—one needle's width displacement of the needle.

When this method doesn't seem to work—you think you have your brackets set and suddenly they don't seem to work anymore—it could be that the wind has changed, but check the directional gyro first. I usually find that I've allowed it to go off, and when I reset it and get back on course, my brackets are OK after all. Even if that isn't the problem, it could be that you've waited too long to start the correction. This is most likely to happen close to the station. People talk about the needle becoming more "sensitive" closer to the station. Since the distances are so much smaller for a given amount of error in degrees, you can fly many degrees off course in only a few hundreds or even tens of feet, and the needle will move correspondingly fast. It should be clear why the proper way to fly the omni is to make a small correction, but make it quickly. And it should be clear too why you need to consider the distance from the station when you decide how much correction to make.

If you do find that you need additional correction, that your brackets aren't working, try shifting both brackets five degrees in the indicated direction. Flying omni well can be a great satisfaction, but it does require a lot of attention to the needle and to the heading and it presents a special traffic hazard in busy airspace, especially close to an airport or directly over the VOR transmitter, so be very careful in your practicing. I've adopted the habit of flying at off-beat altitudes—1700, 2200, and 2700 feet—figuring that most people are trying to hold 1500, 2000, and 2500 feet, and that I can cut down on the chance of collision that way.

Checking the Sensitivity of Your Omni Set

While I've spoken of a 90 degree error, most omni sets don't indicate large errors specifically. Full deflection of the needle is supposed to occur when you have about a 10 degree error, but most sets most of the time are not quite that tight. If you do some experimenting up in the air, it may help you to comprehend just what the deflection means. Tune in a station and turn the course selector until the needle is centered. Note the course indication. Now turn the selector slowly, until the needle is just fully deflected. Note the new course indication. The difference between the two course indications is the error in degrees that your set is indicating at full needle deflection. The sensitivity of the needle in terms of degrees may vary not only with your distance from the station, but also between one station and another, and between one radio and another.

Identifying the VOR Station

It is important to be certain that you have tuned in the correct station and that it is operating normally. The identifying code should be checked. If the station should suddenly malfunction, the code would be removed from the air automatically. The loss of the identifier lets you know that any signal you're receiving will be unreliable. Correct identification of the code gives most people some trouble; that's why a lot of stations now have voice identification as well. Here is a method of checking the code that may help. First, copy the dots and dashes you should hear on the margin of the chart, or on your flight log, large enough to see them easily. Say them over to yourself, letter by letter, and fix in your mind the rhythm you expect to hear. For example, at North Philadelphia the code is PNE, the Morse is • — — • — • • and the rhythm is dit d-a-h d-a-h dit, d-a-h dit, dit. After you know what it ought to sound like, listen to the code. But listen to and match only the first letter of the three that are transmitted. When you are sure that the first letter matches your expected rhythm, listen to the second. When you are sure both the first and second match, listen to the last. If you are doubtful about the match when you use this method, you probably don't have the right station. Double-check the frequency.

Using VOR for Orientation

On flight tests I've found that most pilots have trouble orienting themselves using omni, even after they've found and correctly identified appropriate stations to use (and that in itself may be hard to do if you're really lost and in an unfamiliar area). There are several things that make it harder than it should be. The most common error is not drawing accurate straight lines on the chart, so the point where the lines cross on the chart isn't very near the point where the radials themselves really cross. A second error is failing to draw a long enough line from the station.

These errors are easily corrected, but there is another problem that still remains. You're trying to find your position by drawing two lines-of-position. But the airplane is moving all the time. When you find one radial and draw the line and then begin to work on the other, the airplane will probably be flying away from the first line. Then, by the time you have the second LOP drawn on the chart, you are no longer over the first one and the intersection of the two is not where you really are.

There are two methods you can use to avoid this difficulty. If you are over a good visual landmark, and you hope to be able to identify it on the chart after you get the two LOP's drawn and have the fix that they'll give you, then it would be wise to circle the landmark and stay in the same general area. If you see no particularly good landmarks, it may be a better idea to turn to a heading that matches the radial from the first station. Fly that heading while you tune in the second station and draw the second LOP. Then the intersection of the two should indicate pretty accurately where you are at that moment.

Some Words of Caution

Omni is not absolutely precise. There may be error in your set, but even if it is unusually accurate, there may be unreported error in the orientation of the transmitter on the ground. If you are using stations within forty miles of your position and if you have adequate altitude, and if your omni set is well calibrated, you can expect to be within two or three miles of the fix you get from your intersection of radials. This is pretty close, and should be enough to orient you anew. But notice that a three-mile error in an unknown direction from your fix gives you a circle of more than twenty-seven square miles! If you are trying to use an omni fix to find a grass field, and you don't use any terrain features at all, this degree of accuracy won't make it easy. I've seen it take an experienced pilot as long as forty-five minutes to find Flying W using

VOR radials when he didn't have a Sectional chart aboard and wasn't familiar with the field.

ADF

There is increasing use of ADF now that there are digitally tuned sets. It has the advantage that you can use not only low frequency beacons intended for navigation but also any commercial radio station tower. Using ADF is not too hard if you want to fly to the transmitter, but flying a specific course, especially away from the transmitter, is much harder than flying an omni radial. When you fly omni it doesn't matter whether your compass is correct or not, because you use whatever heading is necessary to keep the needle centered and that gives you the course. ADF is different; all it shows is where the station is in relation to the nose. If you don't know what direction the nose is pointing, you don't know what course you're on to or from the station. That means you must have an accurate compass and make accurate calculations if you want to fly a specific course. If you're going to use ADF, you'll need a lot of good dual instruction seeing how it works.

As long as you don't allow the use of radio navigation to detract from your ability to keep a watchful eye on everything that's going on outside the airplane—weather, traffic, the ground going by—then use it with my blessing. It can make precision flying and meticulous planning even more rewarding.

Radio Communication

Like radio navigation and almost everything else we use in aviation today, radio communication continues to develop, becoming both more sophisticated and more widely used. It too can contribute to both safety and efficiency. At both controlled and uncontrolled fields this is especially true when the ground operator and the pilot are working together. Unfortunately, there is a kind of Us-and-Them attitude in too many pilots and controllers. This shouldn't be; we are all on the same side! We all want each flight we deal with to be both safe and efficient.

➞ *THE PROBLEM OF SHARED RESPONSIBILITY.* The relationship between the pilot and the controller will work best if each of us takes responsibility for what happens. In that way it is like the student-instructor relationship, which is most satisfactory when each person, both student and instructor, retains an inner conviction that it is he who is ultimately responsible for the way the student flies. Sometimes this results in misunderstanding—the student feels he's not being allowed to make his own decisions, or the instructor feels a lack of trust and confidence on the part of the student—but this is less dangerous than the opposite situation where nobody feels responsible, where each blames errors and difficulties on the other.

➞ *LOSS OF THE "ATTITUDE-OF-COMMAND."* The unfortunate thing is that when two people share responsibility for something, one of them tends eventually to relinquish some of his responsibility and to depend upon the other. Controllers are professionals—all of them. They are carefully trained and supervised and upgraded as equipment and procedures change. There are many pilots who are amateurs, or semiprofessionals—less carefully trained, hardly supervised at all, seldom upgraded. One guess who it is who begins to depend on the other. The curious thing is that the tendency to transfer increasing responsibility to the controller is apparent in the ranks of professional pilots as well. The widely discussed airline accident on an approach to Dulles is clear evidence of that. It is and always has been the pilot's responsibility to know where he is and to know what altitude is safe, especially when he has been cleared for an approach, as was the case here. Yet there was an attempt to blame the controller for the pilot's failure to maintain a safe altitude. This goes against everything I believe about the responsibility of being a pilot, and I was shocked and dismayed by this evidence of the deterioration of the attitude-of-command even among professional pilots. If we attempt to blame our failures to fulfill our responsibilities on the controller who fails to warn us of our errors, we can expect that controllers will become more and more authoritarian and we pilots will become increasingly restricted.

➛ *WORKING WITH THE CONTROLLER.* The best controller to work with is the one who is also, or has been, an amateur or semiprofessional pilot himself. He knows what it's like to be flying; he knows what a pilot sees and how he thinks, so he knows how best to work with you. You, on the other hand, will have no opportunity to become an amateur or semiprofessional controller and so find out at first hand what it's like on the other side of the fence. But you can do your share of the job better if you try to see the picture the way the controller sees it.

➛ *DEVELOPING AIR EARS.* I know this is asking a lot. Right now you probably don't understand a lot of what you hear on the radio, and when you push the mike button it may seem to disconnect your brain from your mouth, leaving you babbling or speechless. But, believe it or not, in time you will get air ears as well as air eyes. And there are some painless ways you can practice and develop this knack a little faster.

Records and tapes are good for getting familiar with radio procedures, though some of what you hear on most records isn't very realistic. Some tapes leave blank time for you to practice answering. Even better is a radio that lets you tune in to aviation frequencies and listen to real conversations. You'll enjoy listening and you'll pick up the jargon painlessly. If you do a lot of airline flying, keep an eye out for a little electronic gadget called a Sky Spy. It's the size of a fountain pen, has a tiny earplug like a transistor radio, and with it you can hear the transmissions made from the cockpit of the plane you're in. It is fascinating. In fact, it can get altogether too exciting—maybe you'd rather not know that your pilot is asking for a routing around severe thunderstorms, or is reporting to the tower that the landing gear is stuck!

If you have access to an ATC simulator—maybe a friend of yours owns one—this is another way to get used to radio communication. There is even a tape for a VFR flight. For practice in developing a good and efficient scan of the instruments, nothing could be better than these simulators. But it isn't like being in an airplane in the air, and too much time in one may tend to make you instrument dependent. The normal cost of renting one doesn't seem to me to be justified for the student who isn't yet a private pilot.

Finally, you can always go and visit the nearest tower and sit quietly and watch and listen. Listen while you're flying too. I find the radio too distracting when I'm giving dual instruction, but your instructor may not feel that way. And perhaps it won't be too distracting when you're out solo, practicing maneuvers. Monitor FSS or nearby towers when you fly cross-country, or when you're practicing VOR navigation. You can make it a little easier by learning the most common words and phrases. There is a whole special radio language, and even experienced pilots may have trouble understanding when nonstandard wording is used.

One day when I was introducing a student to the perils and pleasures of flying at a controlled field, I heard the controller call a four-engine transport that was following a light plane on final. He said, "You're eating up your traffic." The pilot answered, "Say again," and though the controller repeated the phrase several times—meaning that the transport was overtaking the traffic ahead of him, the pilot never did understand him, and eventually was obliged to go around. For my student the experience at this field turned out to be a very poor introduction to radio work as an aid to efficiency. First, the controller consistently used nonstandard phraseology. Also, traffic flow was inefficient because he misidentified airplanes and misjudged speeds and distances. At one time he had three different airplanes doing 360's for spacing in the pattern. He cleared airplanes second in line to land ahead of airplanes that were first. Despite all this, he seemed relaxed and comfortable and quick with his instructions. I decided later that perhaps he was an experienced military controller, newly returned to civilian life, new to the handling of mixed light and heavy, amateur and professional civilian traffic, and unfamiliar with the common general aviation airplanes. The last thing I heard him say that day was in answer to a transport that called requesting a series of practice approaches. Conditions were good VFR; there was no more than a normal amount of traffic, mostly light-plane training. But this remarkable controller said cheerfully, "Please go away. I can't handle you."

➳ *DF STEERS GIVE YOU VALUABLE PRACTICE.* You can get some unpressured practice in using the mike and in responding to hearing your own call sign, by doing some practice DF (Direction Finder) steers. If they aren't too busy, the people in the FSS or the tower are glad to work this procedure with you because they need to get a certain amount of practice themselves. Ask your local FSS what facility has DF equipment in your area, and find out when they'll be most likely to be free to work with you. Get up in the air and call the facility. When you ask for a practice DF steer, they may ask if you want to declare an emergency. That may mean they're too busy for mere practice. They know, however, that lost pilots don't always admit their problem unless they're asked, and no matter how busy they are they drop everything to work a DF steer if a lost pilot really needs one. If they aren't too busy, they will ask you to transmit for five seconds (maybe ten the first time) for a steer. You key the mike, give your call sign and a slow count: one, two, three, four, five. Then they'll give you a heading. You turn to it. While you're at it this might be a good time to practice flying without the directional gyro. Every few minutes they'll call and ask you to transmit again, and they'll give you new headings if they find that you are drifting. The series of steers should bring you directly over their facility, and they'll tell you when you pass it. This is a way to get a lot of radio transmissions without having to worry about traffic in a pattern, or about your own approach and landing. Not to mention that you'll be getting practice in using a valuable emergency procedure.

The Basics of Radio Communication

Radio communication differs from both face-to-face and ordinary telephone conversation in one important respect. When you press the mike button to transmit, your receiver is shut off automatically. Not only do you not hear what the other fellow may be saying, you don't even know if he's talking; and if he is talking he doesn't know that you're talking too. So you both may have the mistaken impression that the other has heard what you've said. This can lead to a lot of confusion, and that's why we use "over" which essentially means, "OK, I've finished talking and I'll be listening for your reply." After a little experience you can tell by the tone of voice and by the content of the transmission whether or not the speaker is finished, so in fact "over" is usually unnecessary, and is often not used.

Let's go through the basics of a radio communication. Before you make the call, make very sure that you have the right frequency, that you have the radio turned on, that your transmitter switch—if any—is set for the right radio, that you have the audio selector switch set to speaker or phones, whichever you're using, and that you have the squelch set properly and the volume turned up far enough so you'll be able to hear the answer when it comes. Then listen carefully before transmitting and for long enough to be sure you aren't breaking into the middle of someone else's conversation. Before you key the mike, plan exactly what you're going to say. If you stop halfway because you can't remember what you were going to say next, take your finger off the mike button, in case the controller starts to answer you. It wouldn't hurt to rehearse your whole transmission once or twice. When I began to use the radio I suffered from terrible "mike fright." I used to write down what I was going to say on the margin of the chart and just read it right off to the tower.

The first step is to get the attention of the person you want to talk to, and all you usually say on the initial call is 1) who you're calling and 2) who you are (your entire designation —type of plane and full registration number). When he answers saying, "Go ahead," you repeat your designation, customarily only the last three characters—digits and/or letters —and tell him 3) where you are and 4) what you want. Sometimes it is reasonable to combine the first two transmissions, giving your position and your request in the first call, but my experience has been that the controller, who may be busy with something else, usually doesn't get everything the first time if I throw it all at him without warning.

➻ *GIVING YOUR POSITION.* Identifying your position efficiently is a skill that has to be worked at. If you are really over some large, easily-identified landmark that the controller will know, then by all means tell him. But too many people use major landmarks when they aren't anywhere near them. And too many people use terms that are too vague and general like "over the river" or "over the turnpike." If you are willing to circle over the landmark you plan to give, in the event that the controller happens to be tied up when you reach that point (as I did when I wrote everything down on the chart), that's fine. If you want to be more efficient, translate the position over the landmark into a distance and direction *from* the field, and be accurate within 30 degrees, using compass directions—east, east-southeast, southeast, etc. The advantage of using this method instead of naming a landmark, especially if you're approaching the facility, is that you can use the same direction and just cut down the distance a little if the controller is busy and you have to wait for the frequency to clear as you continue to fly toward the field.

➻ *ANSWERING HIS ANSWER.* The tower controller will now come back and repeat what he has understood you to say, and will give you landing information and another point from which you are to call him again. If you notice that he has misunderstood anything you said, whether your number, your position, or your desires, let him know now, by repeating it correctly. And if you missed any part of what he said, ask him to say again whatever you didn't hear. If you got it all and you can do as he requests, you must acknowledge his transmission. The pilot almost always has the last word on the radio, so don't hang up the mike after you've told him what you want. You still have to let him know you've heard him after he answers you.

The full terminology for that acknowledgment goes "Roger"—meaning you've heard him clearly—"wilco"—

meaning you will comply with his instructions—followed by your designation. But "wilco" is almost never used. The controller will understand "Roger" to include your willingness to comply. So don't "Roger" an instruction you won't be able to follow. If he asks you, for example, to report passing a point that you aren't familiar with—some numbered highway or named institution or building—let him know you're not familiar with the area. He'll reword his instruction in terms of distance from the field. The controller will take for granted that you mean both "Roger" and "wilco" if you omit both and answer only with your designation. Some pilots, however, get in the habit of saying "Roger" without their numbers. If you're going to leave anything out of your acknowledgment, don't let it be your numbers, because occasionally the wrong plane answers a clearance. If the controller only hears "Roger," he won't know it when it's the wrong plane, and he can't correct the confusion immediately.

Help Yourself and the Controller Too

There are some simple things you can do that will make it easier for you and for everyone else. Listen ahead of time to get a clear picture of how the traffic is flowing around the airport. Look for the traffic that may be of concern to you. If you see traffic you've been told to follow, you can help by letting the controller know that you "have him in sight." Listen all the time for your number, in case the controller initiates a call to you.

You can make it even easier on yourself the first few times you go to controlled fields by telling the controller that you want to circle above the field before entering traffic. This will give you a chance to see what's going on and you won't have to deal with straight-in approaches or an abbreviated right-hand pattern, both of which can be very confusing. The controller will probably approve this request and may give you a minimum altitude above which to circle, and then he'll probably say, "Report downwind." So you may circle, check the field layout, and watch the traffic. When you're ready you can let down and enter the pattern and let him know you're there. Keep listening, because he may call while you're circling or letting down and ask your position. If you aren't prepared for that question, it can be a surprise and may leave you momentarily tongue-tied. In that case tell him, "Stand by," while you figure out where you are—one mile north, over the south side of the field, letting down, or whatever. One poor fellow on a flight test with an FAA inspector was so undone by the whole thing that when the controller said, "What is your position?" he answered, "I'm a clerk in the A & P."

Radar: Uses and Abuses

The use of radar needs some special attention. It too can make life safer and easier in the air, but like omni and radio communication it too often doesn't. We'll talk just about VFR use, and not at the most sophisticated levels. You must realize that in general the radar controller's principal job is to deal with IFR traffic. He can handle you VFR only on a workload-permitting basis. Approach and departure controllers usually (but not always) are looking at a radar scope, and at less busy facilities they're glad to give you traffic information. And a lot of major facilities today do have a controller with his own frequency who is there specifically to handle itinerant VFR traffic and give traffic advisories. Find out what the situation is in your local area.

While we're on the subject, radar is not Big Brother in the Sky. It cannot separate all traffic. Too many pilots believe that it can. I was once rendered speechless when a private pilot told me that he had flown along the coast of Long Island, right across the approach to Kennedy's most commonly used runway, at a low altitude in very marginal weather. He went on to say, "Why not? I know they aren't going to let one of those big ones hit me." What he didn't realize was that his small plane, not transponder equipped, was probably completely invisible on the radar scope. If a radar controller says he has you, and he's giving you traffic, you may have a nice secure feeling. But it is false security. He can't tell you about what he doesn't see, and you will notice that people often call and are told by the controller that he doesn't have them on the screen.

This is very important. Radar can keep the IFR planes from conflicting with each other (most of the time); but it absolutely cannot monitor all the traffic. At one time I heard blind calls being made from the ground asking all VFR traffic in an area near New York to please turn off their transponders. There were so many transponder blips that the controllers couldn't distinguish the traffic they were trying to work with. Now the equipment is more sophisticated. We have a lot of special codes, even separating VFR at different altitudes. But severe weather between you and the radar may make you invisible, and besides the controller is human too. So don't depend on him to keep you safe in the air. It's not his life but yours up in that airplane, and it's in your hands, so make good use of the best safety device in the cockpit: a swiveling knob with two fine optic devices—your head and open eyes.

Knowing Your Rights and Responsibilities

Approach and departure controllers talk with authority, as if they had jurisdiction over everything you do. But actually, when you are in VFR conditions, and are not in an airport traffic area or a TCA (Terminal Control Area), they cannot tell you what to do. If you refuse their instructions you may find it harder to get sequenced into the flow of traffic the tower is handling. But if you prefer to do your own navigating toward the airport you can say so, and do it. Radar vectors can be helpful, but they sometimes take you the long way around.

➞ *THE CONTROL ZONE AND AIRPORT TRAFFIC AREA.* Controlled airspace is an area of knowledge that's confusing to almost everyone at first. Your instructor will go over the various kinds of controlled space with you, but there are a few points I'd like to make that may help you make more sense of it. First, controlled space is designed for one reason—to provide safe separation between different IFR traffic and between IFR traffic and VFR traffic. If flight conditions are actually VFR, most controlled space has no effect on your VFR flight. Exceptions are the TCA's around high density terminal areas and high altitude positive control space. I won't go into those here. Most of the confusion for VFR pilots centers around Control Zones and Airport Traffic Areas.

You can find out exactly what these are by studying the regulations and the *Airman's Information Manual.* And you'll hear a lot about them in ground school too. But here let's just look at the simple common sense of them. First, the Control Zone: The rules concerning a Control Zone establish higher VFR minimums than those outside of Control Zones or other controlled space. The reason the Control Zone is there is to protect IFR traffic from conflict with VFR traffic (or other IFR traffic) in marginal weather conditions. If the visibility is at least three miles and the ceiling is at least 1000 feet, and if you (the VFR traffic) maintain 500 feet under clouds, as the rules say you must, then the see-and-be-seen principle should be enough protection. Obviously the Control Zone has to start at the ground, unlike the Transition Areas, because it surrounds an airport and the traffic involved is going to be starting from or heading for the ground. Although many of the Control Zones surround tower-controlled fields (not all—some surround airports served only by the FSS), the Control Zone itself indicates nothing about communications requirements. Flight within a Control Zone is restricted only when weather is below the legal minimums. Aside from that, no clearance is required for VFR traffic desiring to fly within that space.

But then there is the Airport Traffic Area. This is a different story. In the first place, it is not shown directly anywhere on the chart. You have to know that an Airport Traffic Area exists wherever and whenever a tower is in operation. It is a cylinder of airspace, centered on the airport, with a radius of five miles and extending up to 3000 feet. There are no rules that spell out weather minimums within an Airport Traffic Area. Most of them are within Control Zones, so of course Control Zone rules apply. But Airport Traffic Areas do have special rules that have to do with communication. Except under certain conditions listed in the FAR's (Federal Aviation Regulations), you must establish communication with the tower and obtain clearance before entering this airspace. Refusal by the tower to give a clearance is supposed to be based entirely upon traffic considerations. That means that technically a tower can't refuse you a clearance, even if the weather is zero-zero, provided they have no conflicting traffic. Which, please notice, puts the responsibility squarely on the pilot's shoulders, where it should be.

➟ *SPECIAL VFR.* There is one special procedure available to the pilot that makes this whole business even more confusing. That's the Special VFR Clearance which is available at some controlled fields. I feel strongly that students and new private pilots have no business flying with this clearance, so I'm not going to give you tips on how to use it effectively. If it happens that there are special local circumstances—ground haze, or low clouds over flat terrain with good visibility underneath them, for example—which would justify your using this procedure, let your instructor go over it with you, because it needs to be discussed in terms of local conditions and landmarks.

Briefly, what a Special VFR Clearance does is temporarily reduce the legal weather in the Control Zone—just for you and just for the specific clearance you are given—to visibility of one mile and flight clear of clouds. An altitude, or an altitude limit, will be assigned to you, and you'll be expected to comply with the clearance as if it were an instrument clearance. But it *does not* permit you to fly IFR. As a student or new pilot, you just don't have enough experience to handle flight safely in these minimum conditions.

The one best training aid you can find for radio work is a real live controller who is also a pilot. Maybe your instructor, or the Fixed Base Operator where you fly, can get someone who meets these qualifications to come some evening and talk to a group of students and licensed pilots. If so, he can probably clear up a lot of confusion. You'll find that, with practice, using the radio will not only contribute to safety and efficiency but will turn out to be fun too. Don't let yourself be buffaloed by the crisp, impersonal, professional voice on the speaker. It's coming from an ordinary human being like you. You can enjoy working with him to keep everything moving smoothly. You may even get to where you won't feel you always have to stick to the cold official phrases. You won't get exactly chatty, I hope; but you'll manage a "please" or a "thank you," once in a while, or even, under appropriate circumstances, "Merry Christmas," "Happy Easter," or just "So long, have a good day."

12. Those Post–Cross-Country Blues

There are three distinct points in your training when you're likely to get discouraged. The first comes after you've passed through the stage of being afraid you're going to be soloed before you're ready, and you start to be afraid you'll never be soloed, ready or not. The second comes when you wonder whether you'll ever be able to spot railroads and identify towns and do it while you're also holding heading and altitude, not to mention talking on the radio. The third comes after you've finished your solo cross-country and you're practicing maneuvers and waiting for your instructor to say, "OK, you're ready." Let's look at that last blue period and see what you can do about it.

When your instructor signs you off for solo cross-country he is really saying that now you are beginning to be a pilot, and that's how you feel too. It may not be as thrilling as the very first time up in the air by yourself, but in a way it means even more. Going around the pattern, you were under your instructor's wing, even though you were in the air alone. But now, when you leave the pattern for a cross-country flight, you are really on your own. It is no longer enough to be a good student. Now it is important that you also be a good pilot and that means being responsible for where you are going, both literally and figuratively.

It has been traditional, once a pilot begins doing cross-country work, for him to do a lot of solo flying with very little dual instruction. People call it "building time," as if more skill comes automatically with more experience in the air. It doesn't. A lot of that time may be spent practicing mistakes, and then when the student finally finishes all his solo cross-countries and comes back for dual in preparation for the flight test, his flying is worse than it was when he first soloed. Now he not only has to learn what is right, he also has to *un*learn a lot of bad habits, and that will take extra time. So the time he was flying alone has been worse than wasted, it has actually set him back. After building his confidence by completing the cross-country work successfully, he will be doubly disappointed to discover that he still has a long way to go.

Making Solo Practice Count

How can you fly better and build time too? Try this. Don't fly the ten required hours of solo cross-country in a solid block. Instead, mix it up with both dual instruction and solo practice covering airwork, special takeoff and landing techniques, instrument work, and night flying. And make all your solo practice count! Most instructors don't supervise this advanced solo work very closely, so it will be even more important than usual for you to plan these flights very carefully for yourself. Even if your instructor does assign you specific maneuvers to be practiced and reviews the criteria you should be trying to meet, you can, as always, make your time more valuable by adding some ground time to it. Brief yourself before each solo flight, thinking through what you plan to do. Don't try to cover too many different maneuvers on one flight, and if your plan turns into a confusing blur after you get in the air, write out an outline for yourself and take it along. Continue what should be a habit by now: Debrief yourself after the flight, thinking through what you did. And write down what you want to work on next time. I once had a student who laid out every solo flight for himself minute by minute. When he had his forty hours he wasn't just barely passing—he was flying unusually well. He was the best student I've ever had, not because he was the most apt —he wasn't—but because he worked at it so hard and so efficiently.

Now that you have more experience, you're less tense when you fly, and so less quickly fatigued. But the shorter flight sessions will still be more valuable to you per minute bought and paid for. A half-hour session of dual on stalls, followed by debriefing over a cup of coffee, and then a half-hour session of solo practice on the same maneuvers will give you very nearly the same benefit you would get from twice as much flight time in the form of two hour-long sessions.

A very useful aid to you at this point is the FAA *Flight Test Guide.* This tells you exactly what you'll have to do and what the criteria for passing performance are. There are the ground reference maneuvers to practice, and the stalls and slow flight. Always there are normal takeoffs and landings. And now I can't put off any longer the discussion of the various special takeoff and landing techniques that your instructor introduced you to before your first solo cross-country.

Special Takeoff and Landing Techniques

I ask myself why I've been avoiding this discussion. In the back of my mind I've been thinking that these techniques can be difficult enough without my telling you one thing and your instructor telling you another, and some of my ideas are nonstandard or controversial. Besides, this is an area where "appropriate" is really at the heart of it! Though I do have preferred methods, the final choice has to depend not only on the airplane, but also on the field you're using and on your own general aptitude. I can't determine all that; so, if my conscience would let me, I'd be happy to leave the whole business up to your instructor.

But I think I may be able to contribute something useful to you, so here we go! The emphasis on special types of takeoffs and landings, and even the recommended methods of doing them, has changed through the years like so much else. When we were learning to fly out of grass fields, conditions were a lot more variable. There were times of the year when those fields were bound to be soft, so soft-field takeoffs and landings were a natural part of our experience. A lot of those fields were pretty short too, and had trees around them. So short-field takeoffs and landings were also encountered in the natural course of events. Crosswinds, on the contrary, are probably more of a problem now. Although our landing speeds were lower and we set our crosswind limits correspondingly lower, we could and did adjust the landing and takeoff direction across the grass so we were usually headed approximately into the wind. You just can't do that very well when you're using a paved strip.

➳ *WHAT WE MEAN BY SOFT, ROUGH, OR SHORT.* Before I go any further we'll need to define a few terms. Modern pilots use the word "soft" to refer to any field that isn't hard-surfaced. That's not what it means in reference to soft-field takeoffs and landings. A good, firm, sod field may not be hard-surfaced, but it isn't "soft" either. Soft means a surface condition that causes a high amount of "rolling drag"—drag due to the friction of the wheels rather than to the flight characteristics of the wing. This might be caused by mud, or by very wet and soggy sod, such as you are most likely to find in the spring. You'd get the same effect on a sandy beach, or in high grass, or in snow. It's true that even hard, firm sod feels different from a smooth hard surface and reduces takeoff performance somewhat, but it doesn't require any special technique. Another thing many modern pilots seem to think is that all grass fields are rough fields. Sod isn't necessarily rough. "Rough" means really bumpy—bumpy enough so you want to keep the weight off the nosewheel as much as you can—bumpy enough to launch you into the air before you really have flying speed.

In a sense "short" is defined by the figures in the airplane manual. It's a relative term, of course. In some trainers you could take off, climb straight ahead to altitude, descend and land, and do it all again, on a runway that would be too short for a big jet. How short is short has something to do, too, with field conditions and with whether you are trying to land or to take off. When I say this, what I'm thinking of is a field of normal length with a soft surface. Your approach and landing in this situation wouldn't require short-field technique, but your takeoff roll might be so long that your climb after takeoff would call for maximum-angle-of-climb speed. Another circumstance that changes a normal length into a short field, in terms of performance, is the density-altitude. So we'll define "short" this way: Short refers to any field where a combination of runway conditions, obstacles, aircraft loading, and atmospheric conditions make maximum performance desirable. I'd like to point out too that you won't do quite the same thing when the landing area is short and bounded by trees, as you will when it's shorter still but bounded by ditches.

Takeoffs and Climbs

➸ *SOFT AND ROUGH FIELDS.* I want to discuss the different takeoff and climb techniques before I talk about landings. I think it may help to compare the different methods. And since we're talking about takeoffs first, we can start with taxying out when a field may be soft. Let's assume you flew into a grass field a couple of days ago and tied down there. You visited friends, or went fishing, or whatever other excuse you had figured out to justify the flight. Yesterday and last night it rained long and hard. Today you plan to leave. Start by asking the local pilots what the field is like when it's wet—are there any spots that usually get especially soft, or is there an area that's dependably firm? Watch out for the fellow who says casually, "Oh, it might be a little soggy, but you won't have any trouble." It may be true, but on the other hand remember he's going to be sitting comfortably in the office, not trying to pull the airplane out of the mud, in case he's wrong.

Next, take the time to walk out on the field and look at it yourself. Suppose you find it is soft, but not actually boggy. You find a good firm spot where you can safely do your run-up, and the wind is straight down the length of the field, so you elect to go. Now you have to taxi out. It will always take less power to keep moving than to start moving, so once you start don't let the plane stop. And keep the wheel well back to keep the nosewheel light. You'll probably learn this technique on a hard surface, pretending it is soft, and it will be hard to be convinced that it's important. Believe me, it is!

If you feel yourself slowing down, use more power right away, even full power if it's necessary momentarily. If you're accustomed to a hard surface you'll be amazed at how much power this taxying may take. If the airplane does come to a stop, don't try to pull it out of the mud with power because you'll only dig the nosewheel in deeper.

Set extra flaps; you want to get the airplane as light as possible at the lowest possible speeds and the flaps will help. For example, use 20 degrees in the Cessna, 25 degrees in the Cherokee. Don't worry about the drag that flaps cause—it isn't much at the low end of the speed range, and the extra lift is more important now. You'll want to taxi away from your firm run-up area and start your takeoff roll without letting the airplane stop, so be sure everything is set including the directional gyro. You'll begin the roll with the wheel all the way back, but as you pick up speed you may find the nose going too high. If so, just ease forward on the wheel a little.

The most common error in the performance of this takeoff comes just after liftoff. Once you break ground and the surface drag isn't a factor any more, the airplane will accelerate abruptly. The extra speed may tend to pitch the nose up. But you are lifting off at just barely above a stall, so you want to stay in ground effect while you accelerate smoothly and efficiently to a normal climb. Ease the nose gently down toward the proper climb attitude. (Of course, if you have a heavy load or a high density-altitude, that attitude—the "proper climb attitude"—will be more nose-down than usual.) Only after you have a good stable climb and at least fifty feet of altitude (estimated as you are looking outside) should you reduce the flaps to the setting appropriate for a normal climb.

The second common error is one that shows up in the short-field takeoffs too. There is a tendency to let the airplane climb with the left wing down, or even to let it turn to the left. The reason is that at liftoff in these two takeoffs you will feel the effect of the unfamiliar relationship between the power and the airspeed. Lower airspeed relative to the power means that the power effects which cause yaw to the left are more noticeable. At the lower airspeed you'll need even more right rudder in order to keep the airplane climbing most efficiently, i.e., straight ahead with the wings level.

If you ever have to deal with a rough field, use a similar method to that for the soft field—keep the weight off the nosewheel and lift off at a minimum safe speed. The difference will be in taxying, where you'll want to go very slowly. Cross ruts diagonally. After lifting off don't haul back to try to keep the plane from touching the ground again. Let the plane skip across the tops of the rough spots until you've accelerated enough so you rise above them automatically at a safe flying speed.

➳ *SHORT FIELDS.* The short-field conditions assumed for purposes of the flight test demonstration are not likely to be the same as the ones you'll find in real life. On the flight test a smooth, level, hard surface is assumed, and you are supposed to fly so as to clear an imaginary fifty-foot obstacle by fifty feet. This is also how the charts in the airplane manual are set up. In real life, you'll have a hard time finding a hard-surfaced field that's so short as to require maximum performance in a training-type airplane.

Today the fields that are really short and also have obstacles remain unpaved. It just isn't economically feasible to pave a field that will be too short, even when paved, to handle high-performance airplanes. The result is that you may learn the short-field techniques, as you did the soft-field techniques, through the use of Let's Pretend; and again, it just won't be the same. When all you can see in front of you is hundreds of feet of runway and open sky above it, it's easy enough to wait for the proper liftoff speed, and to lower the nose as necessary to maintain your best climb. But it's quite another story when a row of trees is coming rapidly closer and looming rapidly higher in front of you. You have to develop faith that aiming at the thing you want not to hit is really going to work. That's not easy!

Your instructor will go over the airplane manual and the speeds for your airplane. But he may not discuss the three different techniques used for starting the takeoff roll. We used to start as far back on the field as possible and hold the brakes until the throttle was wide open. Another method that's more common on today's hard-surfaced field is to pull out from the run-up area at higher than normal power and go right into the takeoff roll. The advantage is supposed to be that you avoid the short period when the prop is very inefficient as the airplane starts to move. The third method is to use a normal initial acceleration, just being sure, of course, that you start from as near the end of the runway as you can get. The FAA tested the three methods and they say that the normal technique is as effective as the other two. My own experience doesn't refute that, and that is what I recommend. For one thing, it has the advantage that you will be doing what comes most naturally at a time when you may be under stress.

The next matter is the question of when and how to lift off. The FAA recommends a positive rotation at the point when you reach the best-angle-of-climb speed. This method is OK or even desirable for high performance airplanes, especially multi-engine. But in a light trainer it places too much emphasis on "flying by the numbers," on looking at the gauges, when you could do as well by feel and also have the advantage of keeping all your attention outside. Furthermore, this method is suited to a smooth, hard surface, which is OK for the flight test, but may not be what you'll want in real life. Your object is to accelerate to the speed for maximum-angle-of-climb using the least possible space. If the field is soft, or rough, it may be more efficient to lift off at a lower speed and do the rest of the accelerating in the air.

It was in connection with short-field takeoffs that I first realized the FAA could be wrong—even about the application of aerodynamic principles to the flight of light planes. I've been a skeptic ever since. They used to assure us—and I, in turn, assured my students—that while flaps might give us an advantage in terms of takeoff roll, that advantage would be lost by the time we got to the problem of clearing an obstacle. So, of course, we should never use flaps for a short-field takeoff with obstacles.

At Flying W in New Jersey there are tall trees close to one side of the runway. I got curious about just what difference flaps would make, so I began to experiment with both Cessna 150's and Cherokees, watching alongside to see just where I was when I topped the trees. I found that in the Cessna about 10 degrees of flaps or a little more produced the best performance, and that 25 degrees (two notches) made a distinct improvement in the Cherokee performance. The effect in the Cherokee was so striking that I wrote Piper about it. I had a letter back in which they said, yes, they knew that. And a year or two later their airplane manual suggested 25 degree flaps for the short-field takeoff. Of course, different wing designs will behave differently, and in the absence of better information you'll do best to go by what the airplane manual recommends.

I've already said that it's important to use enough right rudder in these climbs. I'd like to suggest that you go out and see that for yourself. At altitude, set up a climb at the speed for maximum-angle-of-climb. Hold the nose straight, as you should do, with rudder. Now watch and see what the vertical speed settles down at. Level out, go back down to the original altitude, and set up the climb again at the same speed. But this time hold the airplane straight with the aileron. The difference in vertical speed may surprise you.

➞ *MORE ABOUT ATTITUDE AND FEEL.* Keep in mind that the nose attitude will not be the same every time you do this. Not only will load affect it directly, as well as altitude and temperature and humidity, but the power you are getting will affect it, and that will be affected by atmospheric conditions too. The fact is that this variation in nose attitude could be raised as an argument by anyone who doesn't like my emphasis on flying by attitude as opposed to flying by the gauges. But remember that along with that emphasis on attitude has been a lot of discussion of getting the feel of the airplane. Again, the method I like is more appropriate for the lighter airplanes, the airplanes in which there is a lot of variation in the feel and response of the controls. Mushiness, or a too-slow response to control use, is there to be felt when the airplane is too slow, no matter what the attitude and airspeed seem to be telling you. And incidentally, because of the emphasis on the speed for best-angle-of-climb, you may still be tempted to watch the indicator too much. If you do, you're likely to wander somewhat in your heading (losing performance when you do), or to let the pitch angle change. Since airspeed can only change *after* attitude changes, if you're using the indicator as your primary reference for pitch you'll be a little behind all the time.

➞ *WHEN TO ABORT A TAKEOFF.* One more point about short-field takeoffs. Choose a point on the ground at which you will abort the takeoff if you haven't yet lifted into the air. As mentioned in connection with the go-around point, you can't really tell how much runway remains when you're on the ground. So decide just how far you're going to go while you wait for acceleration to liftoff speed. If you reach the abort-takeoff point before you reach the proper speed for liftoff, close the throttle and abort the takeoff. Do a couple of aborted takeoffs dual. The sudden reduction in power may result in a swerve to the right and you may be glad of the practice with the instructor there to help the first time. And, of course, an aborted takeoff may be called for due to power irregularity, a door not properly shut, or some other unexpected happening, so this practice may be valuable in other than short-field takeoff situations. It isn't necessary to stand on the brakes when you're simulating an aborted takeoff. That would be hard on brakes and tires, and there isn't any doubt but that you'll use the brakes without hesitation if you get into a position where you need them.

➟ *CROSSWIND TAKEOFFS.* There won't be any need to simulate conditions calling for crosswind techniques; these will be a fact of every pilot's life. And we don't have a definition problem here either. A crosswind is a crosswind. But there are problems in this discussion too. How much crosswind is too much crosswind? Of course, this is a matter of what's appropriate again. The usual factors have to be considered—type of airplane and your aptitude and skill level—as well as runway conditions (Is the surface wet, icy, or soft?), and even, indirectly, the part of the country and the actual direction of the wind. These last factors are involved because the amount of gustiness associated with the crosswind should be considered when you're thinking about whether or not to fly. There are parts of the country where almost any wind is likely to be gusty, and others where even quite strong winds blow steady and smooth, and yet others where wind from one direction blows steady and smooth, while wind from another is invariably gusty and rough.

Crosswind component limits are often given in today's airplane manuals, or you can use the rule-of-thumb which sets a limit of 20 percent of the power-off stall speed with wind at a 90 degree angle to the landing direction. But these limits assume steady, smooth crosswinds! Most of the time wind is gusty, and it is generally more gusty the stronger the wind. So don't depend too much on the book figures; allow a good margin for the unexpected. Remember too that wind usually increases during the day. You may go flying early some morning in a pleasant breeze and come back a couple of hours later, or even a half hour later, to find that there's a howling crosswind and a real problem.

The best way to learn to handle crosswinds properly is in a tailwheel airplane: The airplane itself will let you know very quickly with quite impressive gyrations if you're doing it wrong! But, as I keep reminding myself, there's no use yearning for a past that's gone, so we'll assume you're in a typical, tricycle-gear trainer. Again, there will be some difference between the low wing and the high wing. One big advantage of the low-winged plane's low center of gravity is that the airplanes were much less susceptible to wind. The Tri-pacer and the Colt were the Piper airplanes that were replaced by the Cherokee. They were high-winged and "short-coupled"—the wheels were close together—and the airplane was as tippy as a three-legged stool. Piper hoped to eliminate a major source of student and low-time pilot accidents through the new design, and I guess they did. But you never get something for nothing, and the wide-based, low-winged airplane, because it is less affected by poor pilot technique, is also less helpful, giving you fewer cues, if you're trying to develop and maintain really good landing technique. That's why I prefer the high-winged trainers.

Crosswind technique depends, to some extent, on how strong the wind is. In the event of a really strong wind, you'll need to hold the airplane on the ground a little longer than usual and actually complete your takeoff roll with the airplane tilted noticeably into the wind and the downwind wheel off the ground. In a low-winged plane this will probably be impossible, because of the low center of gravity and the wide wheel base. The ground effect will make the airplane very light on the wheels before it has enough airspeed to fly. If you don't keep it as heavy on the windward wheel as you can with aileron, you may find yourself sliding lightly across the runway sideways just before liftoff. However, this takes a really strong wind, so you won't encounter this problem if you impose reasonable crosswind limits on yourself—either those of the manual or the rule-of-thumb 20 percent.

The basic technique is pretty simple—aileron into the wind is almost all there is to it, using whatever rudder is necessary to keep the roll straight. The airplane will have some tendency to "weathercock" or turn into the wind. If the crosswind is from the right, this may mean that you won't need any of the usual right rudder during the takeoff roll and may even need left rudder. There are some planes where heavy right rudder is required during a normal takeoff roll—so much so that adding a weathercocking tendency to the left will exceed the rudder available and result in a ground loop on takeoff. The Swift is one like this. In a strong crosswind the Swift pilot always takes off with the wind from the right offsetting the usual right rudder requirement and assuring adequate control. There will be a small increase in performance in any airplane taking the crosswind from the right instead of the left since control displacement (rudder in this case) creates drag and reduces performance. But aileron will be necessary whichever side the crosswind is on.

You aren't going to climb out continuing to hold that aileron into the wind—you already know that wouldn't be efficient. So you'll have to make a transition from the slip you use during the takeoff roll to a straight climb with a wind correction angle and the wings level—that is, you'll transition from a slip to a crab. If your takeoff roll and initial liftoff are straight down the runway, the airplane is moving sideways relative to the air that it's moving in. If you remember the slips, you remember that if the airplane is left to itself—if the controls are neutralized during a slip—the airplane will tend to turn so as to streamline the fuselage with its motion through the air. Here the turn will be into the wind, establishing just the crab you want for the climb. There is no need to rudder the nose around into the wind, nor to bank and turn. Just level the wings with the aileron as you make a positive rotation at liftoff and the nose will swing into the wind like magic. Of course, once your heading is established, you'll need the same right rudder you always need in the climb. Remember, though, that the windspeed and direction are often different as you get higher, and you may need to make some small (coordinated, of course) turns to keep your track straight.

I hope that adding my voice to the fray will stimulate you to think a lot about these procedures, to ask a lot of questions, and to watch what's happening more carefully than most people do. If so, you *may* actually spend the amount of time and attention these matters deserve, and it will all have been worthwhile.

Landings

➟ *SOFT AND ROUGH FIELDS.* The soft-field landing is the easiest to talk about, because there's no reason that the approach itself should be different from any normal approach. But there is the question of how you're going to know from the air that you need to touch down using a soft-field technique. You'll see such signs as: standing water and wet-looking spots, tall grass waving in the breeze, snow. Whether sand is hard or soft can't necessarily be distinguished from the air. If in doubt, use the soft-field method for touchdown. It may help and it can't do any harm.

The whole point when dealing with a soft surface is to prevent the airplane from nosing over, which will happen if it comes to a stop too suddenly. You might think the nose-wheel would prevent nosing over, but actually the nosewheel presents its own problems. It can dig in and snap off as Kershner points out. And it is the fragility of the nosewheel that requires you to use the same technique for the rough field. Also, when an airplane has tricycle gear, the natural balance of the airplane puts weight on the nosewheel when it's on the ground, and this is just what you don't want in the soft or rough field. What you want is to touch down gently, on the main wheels, and then to keep on moving, with most of the weight still on the mains. If you normally make landings at a full stall with full flaps, then you can use your normal landing here, but add a little power as the airplane floats just off the ground to be sure that you can keep the nose up and that you don't misjudge and drop in. That could be as bad as landing too fast. On a really soft field you will have to add more power after touching down in order to keep the nose up and keep the airplane moving. This could even take full power in extreme conditions. Again, since you'll probably be simulating these and pretending you have surface drag, you can't really get the feel of it. But if there's a sod field nearby where you could do a little taxying in high grass, you can get at least some idea what surface drag is like (beware of overheating your engine, though, and of hitting rocks or posts hidden in the grass). Nose-up, and power enough to keep the plane moving—that's really all there is to it.

➻ *THE SHORT FIELD.* Now we come to a controversial one—the short-field approach and landing. Some points are agreed upon by all. Full flaps should be used, and power should control the descent. But few people are taught a technique that makes proper allowance for human nature and the natural reactions of the normal person. A commonly taught technique is that of watching the spot you want to land on and keeping it in the same spot on the windshield. This method will work if you use it properly—with a fixed flap setting and precise control of airspeed. The problem is the part about precise control of airspeed. I've found that this method seems to encourage the usual tendency to point the nose where you want the airplane to go. Then airspeed control goes out the window and you'll be totally unable to control the touchdown point. What happens is that you unconsciously lower or raise the nose in your effort to keep the landing point exactly where you want to see it in the windshield. The end result is that if you start to get too low, you end up both too low and too slow, while if you start to get too high, you end up both high and fast.

There is a way to set up short-field approaches that makes it easier to establish and maintain good speed control. Set up the configuration for the final approach while you're still on the downwind leg. Set full flaps and slow down *now* to the speed you'll use for your final approach. At this point—slow with full flaps, but flying the normal approach pattern—you'll need the power just to get around the pattern to the runway. So you can hold the airplane high until you have just the slope you want—a little steeper than normal—and then come right on down that slope, adjusting the throttle if you need to either steepen or flatten the descent, but trying for an approach in which the slope (the imaginary flight of stairs) remains the same throughout.

Some instructors teach a method of flying the final approach flat, low, and slow until the obstacle is cleared, and then increasing the descent. This is not good. First, you'll be nose high and won't be able to see well on final if you use this method. Second, you'll be very vulnerable to misjudgment when you increase the descent angle close to the ground, and you'll be likely to drop in really hard. The steady, steeper slope that will clear the obstacle using constant power or very small changes is easier and safer. During dual sessions I tell my students to try to handle the throttle so that I can't hear or don't notice the power changes. This can be done if you adjust the power immediately for any small error you see. But be careful of falling into the error of too little correction and too late, or the opposite error of constant throttle-jockeying—too much, too little, etc.

One point about practicing these: Don't use the numbers on the runway as your point of intended touchdown. I personally have seen three different airplanes lose their nosewheels when the pilot, who was trying for the numbers, misjudged and landed just short of the runway pavement, and the really sad part is that all three were only for practice, done at airports with plenty of runway. So assume that your fifty-foot obstacle is right at the end of the runway, which will put your touchdown well beyond the threshold. The point, after all, is not *what* spot you land on, but whether you can land on *the* spot you're shooting for. Since you'll be slower than normal you should start the flare a little closer to the ground and coordinate power reduction with it. Don't let the power you've been using keep you floating too long. If you do touch down with some power still on, and sometimes you will, get rid of it immediately. Here again there is no need to prove that you know you should use brakes as soon as you're on the ground.

Spend a lot of time on short-field approaches and landings. Go out to the practice area too and practice descents up at altitude with full flaps and partial power at the appropriate speed, so that you know how the airplane sounds and looks and, above all, feels in this configuration. Really look at the response of the airplane when you use aileron, rudder, and elevator and compare these responses to those you get during a normal approach. No one stays good at this technique without regular practice, so don't let yourself get rusty. Even if you never go to a short field, you may find it useful when you're obliged to follow slower traffic in the pattern.

Just for a change you might like to see how short you can make the landing roll if you assume no obstacle, but a ditch instead. The idea then is to get down just off the ground, with full flaps, of course, and slow-fly with plenty of power until the "ditch" goes under you. Chop the power, and you're down, and almost stopped too. Incidentally, how would you land if you had reason to believe you had no brakes? Right! Short-field technique.

➳ *THE CROSSWIND LANDING.* This one inspires impassioned debate wherever VFR pilots gather to one-up each other while they wait for the weather to clear. And don't think all the instrument pilots agree on crosswind landing either. Most, but not all of the disagreement arises out of experience with different types of airplanes, at different types of airports, at different levels of pilot skill. In other words, it's a matter once again of what is appropriate, not of an absolute Best Method.

There are three methods you can use: the crab, the crab-plus-slip, and the slip. In a sense the crab alone is not a bonafide alternative because when you touch the ground you'll need to use the controls as you do in a slip. There are a couple of interesting exceptions. One is the Ercoupe without rudder pedals. The other is any airplane with castoring wheels. A lot of Helio-Couriers were equipped with them, and I knew a Cessna 170 that had them too. These wheels make it possible to be rolling in one direction (like straight down the runway) while you're heading in another direction (like into the wind). This feels weird, looks weird, and handles weird. But it does make a crosswind landing possible without a slip.

We're going to ignore the exceptions where a crab-only method is possible, and talk about the combination method versus the slip method. What I think is best for inexperienced pilots is the slip method. I feel very strongly about this, but my view is not the more popular one. I base my position on hundreds of things I've seen in the way of errors, incidents, and accidents, and on what I know about the way our minds work. I am assuming that as an inexperienced pilot you are flying a light trainer-type airplane in a wind that doesn't exceed 20 percent of the power-off stall speed. If you have trouble holding the runway line using a moderate slip, you probably should be using a slightly higher airspeed—in some airplanes a slip causes small errors in the airspeed indicator.

There are arguments against my view, of course. First, some people point out that most manufacturers recommend against prolonged slips. Largely this has to do with engine cooling, and a slip throughout the final approach won't last long enough to be a problem. It is also true that some manufacturers advise against slips with the flaps extended. I've been told that the reason is that some airplanes have a tendency to pitch uncontrollably when slipping with flaps. I know of no case where this has happened at normal approach speeds. If you have misgivings, go up to altitude and do some cross-control stalls from slips. Some people have always recommended landing without flaps in crosswinds. I think the use of normal flaps is better. Again, I'm perfectly willing to acknowledge that the best method depends on the airplane, wind, pilot, etc. But just because a no-flap method is best for some particular airplane you may someday fly in gale-force winds, is no reason to use that method now.

➞ *THE COMBINATION METHOD.* The thing I don't like about the combination method is simply that it is harder to learn to do correctly. First, you must establish a crab or wind-correction angle when you turn onto final that will keep your position lined up with the runway. The right amount of crab is hard to judge, and the small changes you may need as you descend are even harder to judge, maybe because you don't have the runway where you're accustomed to seeing it—right in front of you—but off slightly to the side, because of your wind correction angle. But the real problem arises at the point of transition. If the transition is begun well out on the final it isn't so bad, but most pilots using this method wait until the last minute during the flare, and then the gyrations are hair-raising. All you're trying to do is set in a slip that will allow you to touch down with the nose pointing straight down the runway, while the airplane is also moving straight down the runway, but this is a lot harder than it sounds.

Imagine the airplane on final, heading a little to the left of the runway because of a crosswind from the left. We'll assume that the track so far has been OK—in line with the runway center line. When you get down to the point where it's time to begin your flare, you want to bring the nose around and line it up. How do you suppose you'll do it?

Because of that old bugaboo, wheel-steering, your first instinct will probably be to turn the aileron the way you want the nose to go—to the right, away from the wind. Even if your turn is coordinated you'll drift during the turn and you'll wind up no longer lined up with the runway. If your turn isn't coordinated, it will take a bit longer and you'll be even farther off your track. So by the time the fuselage is parallel to the runway you'll probably be too far downwind to the right of the runway line to try a landing. You'll have to go around and set it up again.

Well, then, what are you going to try next time? Maybe you'll remember you've heard people say, "Kick it around with the rudder." But think a minute. Do you remember what happened on that first lesson when you used rudder to turn the airplane? Right! The wing on the outside of the turn went up. So if you kick the nose around with the rudder in this case, you'll have to do something about that. You can keep the wings level with the opposite aileron, and this will work a little better, but while you're turning, the wind is still blowing and you're still going to end up to the right of the runway. Of course, you can go around again, but that won't help you do it right. It would be funny if it weren't pathetic, watching people on really windy days come around again and again trying to make the same impossible maneuver.

The point is that the wing has got to go down into the wind *as* the wind correction angle is reduced, not *after*. This transition from crab to slip is really a simple transition from a straight glide to an ordinary forward slip. It looks different because the glide is in air that's moving sideways across the runway. The slip to a landing in a crosswind is usually called a side slip, but remember that all slips, once established, are indistinguishable from each other. What defines the difference between forward and side slips is the relation of the path during the slip to the path before and after the slip.

Since slips today are usually used only when wind is blowing across a landing runway, most pilots don't really understand the distinction between forward and side slips. For that matter, most pilots never did. A forward slip is a slip in which the flight path during the slip is lined up with the flight path before and after the slip. It can be entered by using abrupt aileron, which causes adverse yaw, and then holding the nose on the new heading with the rudder while holding the wing down with continued aileron. The recovery is a matter of timing the release of the aileron and rudder so that the nose swings back to the original heading. In a side slip the track of the airplane during the slip is at an angle to the track before and after the slip. Here the nose is held on the same heading as the original one throughout by timing application of aileron and rudder differently. A little thought will prove to you that a slip may be, at one time, both a side slip and a forward slip. That is, the flight path during the slip may match the path before without matching the path after or vice versa. Practice these in the air until you understand and can see the difference. You'll catch on to crosswind work more easily if you understand slips first.

In this case you want to maintain the same path, so the technique you use is the one that's used to establish a forward slip. Remember how that goes? Roll the wing down rather abruptly, with a fair amount of aileron. This will yaw the nose toward the wing which is rising. In this case, of course, you'll be rolling the windward wing low so the nose will yaw toward the runway just as you want it. Catch it at the end of the yaw and hold the heading with the rudder. When you use this method, the wind has no opportunity to drift you out of line. When the combination method works, this is how it's done. But it's not as easy as it sounds. What makes it so hard is having to turn the wheel opposite to the way you want the nose to go when you make that transition. This isn't easy for anyone, and for a lot of people it seems to be nearly impossible during the landing process.

➻ *THE SLIP METHOD.* When you use the slip method, you will roll out of the turn onto final directly into the slip. If you have misjudged your turn, or have set in too much or too little slip, you have the whole final in which to correct. And you will be doing exactly what you do on every normal final approach, only more so. That is, you will "hold the runway line"—keep the airplane position lined up with the runway—by using aileron; and you will maintain a heading that keeps the fuselage parallel with the runway by using rudder. If you have trouble seeing that, a helpful exercise is to go out (dual) when there is little or no wind and the air is smooth and make several approaches during which you purposely slip the airplane from a position lined up with one side of the runway over to a position lined up with the other side and back to the middle. I've found a good general rule for these small slips is to take out the slip when you're about halfway across to where you want the slip to take you. The airplane slides on over to just about the right place while you're recovering.

I've noticed that a lot of pilots do everything right on the crosswind approach until they are within inches of the ground. At the last minute they seem to be unable to complete the process correctly. What I think creates the difficulty is an unconscious sense that the airplane should be level when it touches the ground. Not true. You want one main wheel to touch down first. Convince yourself that, like it or not, that is really what you want to do. Of course, you may sometimes have to reduce the slip or even level the wings as you approach the ground because of a temporary change in wind speed or direction, but if it happens that you do touch down with neutral aileron, turn the wheel back into the wind after you're on the ground. Hold the nosewheel off with back pressure at least until both main wheels are down and longer if you can. Once you have both main wheels on the ground the wings are level, no matter what you're doing with the aileron, and the airplane no longer has any strong tendency to turn into the wind—so the rudder you needed in the air can be released. Then the nosewheel can straighten out and touch down without any sideward load.

A lot of the gyrations that have brought my heart into my throat have been observed *after* the touchdown in a crosswind landing. What may happen is that the approach has been a little tricky, and the pilot's keyed up and tense. All his attention is on the problem of setting the airplane on the ground more or less in the middle of the runway. Once he's done that he thinks, "Ahhhhh. Thank God! That's done," and he relaxes and stops flying—a minute or two too soon. Remember, the landing isn't finished until you taxi off the runway, and if you don't maintain aileron into the wind and back pressure on the elevator through the landing roll, things can go sour awfully fast.

That matter of keeping back pressure on the elevator is another example of a turn-around in recommended technique. This one took a long time coming! In the mid-fifties, when a lot of pilots were transitioning from tailwheel planes to the new tricycle-gear Cessnas and the Piper Tri-pacer, the current dogma said to get the nosewheel on the ground immediately so you'd have directional control. The big advantage of the nosewheel was supposed to be better visibility and directional control on the ground and nobody could see how you could get this advantage unless you got that wheel right down on the ground.

As you already know, I learned on grass fields, and from people who had a direct stake in the airplanes and so were concerned about operating them without maintenance problems. I was taught to go easy on the nosewheel—keep it light on the ground. So I knew from my everyday experience without even thinking about it that there were no directional control problems if the nosewheel was off the ground. Of course, the nosewheel provides much, much greater ease in taxying. But in maintaining a straight takeoff or landing roll it is not the nosewheel that makes the difference. Rather, it's a matter of where the center of gravity is in relation to the main wheels and the drag of the ground, and the fact that it isn't necessary to raise the tail during takeoff. If you experiment with a high speed taxi, first with the nosewheel on the ground hard, and then with the nosewheel light, you won't have any trouble seeing that directional control is easier when the weight is off the nosewheel.

The FAA finally put out a recommendation about keeping the nosewheel light on the landing roll, referring to what too often happened otherwise as "wheelbarrowing." An FAA inspector once told me that he had won ten dollars when he bet another pilot that one of the next three airplanes to land would touch first on the nosewheel. Actually, he said, two did so. It was probably not coincidental that all three were low-winged planes. Most of these rock forward onto the nosewheel even if some back elevator pressure is maintained unless there are back seat passengers. It is easy for a pilot to get careless about his landing technique without being aware of it—another reason I like the Cessna for training. If you notice that you are suddenly having some trouble with directional control after touchdown, check your pressure on the elevator. You are probably relaxing it too soon.

Gusty Air and Wind Limits

I can't leave the subject of special takeoff and landing techniques without saying something more about gusty air. Gust differential is the crucial thing. Usually the gust differential increases as the wind increases. And it isn't a figure you can get directly from weather reports. Wind reported as "18 gusting to 25" doesn't have a 7-knot gust differential, because 18 is an average, not a minimum. Actually the gust differential—the spread between a lull in the wind and a maximum gust—is about twice the difference that's reported. The rule of thumb is to add half the gust differential to your approach. A better rule is to add enough airspeed so you have the control response you need to maintain the airplane position and attitude. The worse the gusts are, the more extreme and violent will be the attitude changes they'll cause, and the more aileron and rudder response you'll need to make your corrections quickly enough so that a wing-low attitude can't change your heading or position to something you don't want.

Going back to rules of thumb: They say to set a limit of 40 percent of your stall speed for a wind straight down the runway—if the wind is stronger, don't fly. Most trainers used to stall at under 50 mph. That meant a maximum of no more than 20 mph and in this range gusts are not very severe. Today most trainers stall at higher speeds, so if you go by the old rule you run a chance of getting into some pretty rough air. Make a rule for yourself that allows for the airplane you're flying and the kind of winds you have in the area of the country you're flying in. If you go on a long cross-country journey, remember that wind in other areas may be different and may require an adjustment in your rule.

I'm not going to get into a full discussion of very gusty or very strong winds. However, I will describe briefly what I recommend for very strong (and gusty) winds. We all agree that extra airspeed is required. But controversy arises about how to get it—power or attitude? In a strong wind you will soon find yourself much too low if you turn onto final at the altitude you usually use. So it is often true that the stronger the wind, the flatter the approach. However, since most pilots don't allow for the effect of strong wind on their final approaches, they are obliged to carry extra power just to reach the field. And their flat approach means they're flying where the most turbulence is and are very vulnerable to a stall caused by a gust differential.

It is far better to set up an approach that is much higher and steeper than usual, using the extra steepness to give you the extra airspeed you want. Once you are accustomed to the unusually steep angle of descent and the unusual nose-down attitude, you'll become aware that you have *more* control this way than in the flat power approach. A special advantage is that a sudden lull, though it may steepen your descent, is less likely to result in a stall, because your nose-down attitude protects you. This method isn't appropriate for heavy airplanes—it takes them too long to change the flight path at the bottom of the descent. But this is definitely the best method for light single-engine and twin-engine airplanes. Furthermore, contrary to common practice, do use flaps when the wind is strong—you can get a more nose-down attitude without excessive speed and you can get rid of whatever extra airspeed seemed necessary during the descent much more easily when you begin to flare.

The winds that really require this type of approach are stronger than you should be tackling yet. When you're ready, you should get some dual on this technique. The professional pilots I've shown it to have become believers, but it is not widely used, and you may have some trouble finding someone who is familiar with it.

Some Final Remarks About Takeoffs and Landings

One very important thing to remember, and it applies to all landings, not just to crosswind landings, is that a good landing very seldom follows a poor approach. If your whole approach has been uncertain and erratic, you'll be much better off if you go around and try to set it up right. A more general rule is that knowing your own limitations is the key to safe flying. This means that you will think about your own capabilities, not about what the book says, and you will allow some margin between the forecast or observed wind and the crosswind component you're sure you can handle. If worse comes to worst and the wind is definitely beyond what you know you can handle, don't be ashamed to go to an alternate field. He may grumble, but any instructor would rather drive many miles to pick you up than walk a few hundred feet to the end of the runway to look at the crumpled wing tip or snapped-off nosewheel.

If you are *in extremis*—the wind is awful and there's no other field within reach—don't just fling the airplane on the ground trusting to luck to make it right. Keep firmly in your mind that 1) you are the Boss of the airplane; 2) you have the option of using power, either to keep the plane off the ground until it is just where you want it, or to go around and try again; and 3) there will be moments when everything will be just right, even in very strong winds. So if you have the stamina and self-discipline to make each approach with a sense that you're practicing, you will eventually get one that will work all the way to the ground. Just remember *not* to relax when you're on the ground!

I was once asked, "What do you do if you have gusty air (requiring extra airspeed) and a short field too (requiring a slower approach)?" We need more data before we can answer such a question. How gusty? How short? Who's the pilot? What's the airplane? You cannot expect to fly by set rules, looking up formulas to cover each specific problem. No set of rules can cover all the eventualities of life, unless you include a general catch-all rule. The catch-all in aviation, as in almost everything, is: Always use your own best judgment.

13. Finishing Up

You are on the home stretch now and there are still a few things I want to talk about.

Preparing for the Written Test

The chances are there will be a good ground school course offered at the airport where you're flying. One advantage to being a member of a class is that you benefit from the answers to questions you might not have thought of asking. But if it happens that the ground school schedule is impossible for you, you may elect to fulfill the ground school requirement by doing a home study course supervised by your instructor. In either case I recommend that you precede the taking of the FAA written test with the taking of the Acme practice tests. You'll find these advertised in most of the magazines and a lot of airports stock them too. Each book contains three sample tests which are very similar to the actual government tests. Answers are given in the back of the book, and you can also purchase a book of explanations.

Some people just read the questions and learn the correct answers, but there is a better way to use these tests. I'd suggest that you actually take one test without looking up the answers until you've completed it. Since the questions are multiple choice, like the FAA test, you may be tempted to guess the right answers. That's OK on the FAA test where wrong answers don't count against you any more than blanks. But don't guess on this practice test. Instead, leave blank any answer that you aren't sure of. The point here is not to get the best score you can luck out with, but to find out what areas you need to study further. Suppose you make a lucky guess, for instance, on whether a medical certificate is valid for twenty-four calendar months or for twenty-four months. Going over your errors later, you won't realize that this is a point you aren't sure of. When you come to the similar question on the real test, you may guess wrong. After you check your answers on the first test, looking up the answers you left blank and studying the areas you had trouble with, take the second practice test. Go through the same procedure, and ask your instructor for help if you need it. Then take the last test. By now, you should be getting a really good score and your instructor should be ready to approve you to go and take the real test.

Emergencies

Another thing that remains to be discussed is item 15 on the list in Chapter 3: "Anticipate, recognize, and assess unusual, hazardous, or emergency situations, and take the proper corrective action." By emergency situation I mean a situation you didn't expect, one that requires some special action or change of plan and that also entails some hazard. By this definition a situation that would be an emergency for one person—something he didn't expect or plan for that entails some hazard—might not be an emergency for another person who has anticipated the possibility and has worked out a plan for coping with it.

For example, take a night cross-country. John J. Optimist depends entirely on the radio—flying on top of clouds because it's smoother, and not even thinking about possible ground references or airports along his route. He has enough gas to get where he's going, but not much more. He doesn't have a flashlight, or Sectional charts either. If he loses only as little as his navigation radio, poor John has an emergency on his hands.

In contrast Joseph J. Prudent, making the same flight, has checked the chart for the ground references he may be able to use—towers, small towns, highways, lighted airports. He's checked along the way, thirty or forty miles to each side of his course, for lighted fields and for visual brackets that might help him at night. He stays under clouds and he keeps track of his position visually. He has plenty of gas. He has two flashlights, and he doesn't go anywhere, day or night, without Sectional charts. For him the loss of his navigation radio, or even his voice communications as well, won't be an emergency. If he loses his whole electrical system his only problem will be that of being seen by other airplanes, and possibly making a landing at a controlled field if that seems to be his best alternative. Mr. Optimist is likely to have many more emergencies in his flying career than Mr. Prudent. I guess I'm trying to say that emergencies are what you make them. I like to think that the reason I don't have a lot of hair-raising tales to tell about my flying is that I plan and think ahead and try to imagine what could possibly go wrong, and to figure out how I could tell as early as possible that something was the matter and what I could do about it.

Between your instructor, Kershner, FAA publications, and the articles and accident reports in the magazines you should get a pretty good coverage of the various sources of emergencies. I want to talk about a few related points that are often omitted or underemphasized. The first of these is: Unlatch the door before touchdown if you're making a forced landing. The reason for this is that it is possible for the frame to be bent in such a way that you won't be able to get the door open after the airplane stops, and you and your passengers could be trapped in an accident.

Coming to Earth Right Side Up

Here's another thing I feel very strongly about—again because of accidents I've seen or in which I've known the pilot. Maintain control of aircraft attitude until the airplane stops. That's nothing new, I know; that is, everyone will tell you to maintain flying speed, which is simply another way of saying the same thing. But I'd like to discuss this a little. The construction of an airplane is such that if the pilot maintains control of the attitude (by not letting the airplane stall out of control), he can settle into trees, rocks, water, corn, or smooth ground—whatever—nose-up and the airplane is likely to take the brunt of the impact. If the pilot loses control of attitude and comes down on one wing or on the nose, the damage will usually be much greater and the pilot and his passengers are exposed to much greater risk.

That sounds obvious. You'd think everyone would follow the sensible path and be certain to come down with the wings level. But we all feel that if we can save the airplane from damage we can be sure of also saving ourselves. As a result the pilot too often attempts to save an airplane that's just not savable, and in the process he loses the whole thing.

What I'm talking about shows up in two different kinds of

fatal power-loss accidents. The first is the result of attempting to stretch the glide. I remember reading years ago about a mid-air collision between two light planes. One had a wing torn off; the other was only slightly damaged. The first spiraled to earth like the seed of a maple tree and the pilot, remarkably, wasn't badly injured. The second, after gliding a mile and a half, crashed nose-down as the pilot apparently attempted to stretch the glide to get into a parking lot. That pilot died trying to save the airplane.

The other kind of accident happens over and over. In the accident statistics you'll usually find it listed in the stall-spin category, and it's almost always fatal. A pilot takes off and climbs a couple of hundred feet and then the engine quits. He knows perfectly well that he's supposed to go straight ahead from this altitude in this situation, but when he looks straight ahead he sees only rough ground, or bushes, or fences, or something else inhospitable. There's no place to go where there won't be damage to the airplane. His subconscious refuses to face the hassle of an accident. So he tries to turn and get back into the field. Trying to hold altitude in the turn, he stalls and the airplane comes down in a wing-low and/or nose-low attitude. He won't have to face the hassle of this accident, but someone will.

I became particularly sensitized to this kind of accident very early in my career as an instructor. The man who took my place when I left my first aviation job died with his wife in just such an accident only a few months later. He was flying out of the field I knew so well, and I could not understand how an instructor who had doubtless taught many students not to try to turn around in this situation could do it himself. I've been close to many similar accidents, and in every case there was some area in which the airplane could have been crash-landed without likelihood of injury to occupants, but no area where damage to the airplane could be avoided. I've thought about this so much that I am sure that while I may someday damage an airplane I could have saved, I will never die trying to save one I couldn't.

Avoiding Complete Power Loss

The most common causes of forced landings are engine failures caused by one of three things: fuel exhaustion, fuel contamination, or carburetor ice. Engine failure on takeoff is likely to be caused by fuel contamination and is most likely when the airplane has been sitting unused for an extended period with partially filled gas tanks. The forced landing en route is more likely to be caused by fuel exhaustion or carburetor ice. Good judgment and good procedures can easily prevent all three of these difficulties from ever causing you trouble.

The book says that you'll know you have carburetor ice because of the power loss. The problem is that ice may cause only a very gradual loss of power, which you may compensate for by opening the throttle, until finally the build-up is sufficient to make the engine quit. Here is where the habit of flying properly trimmed and with a light touch on the wheel can give you warning in time to prevent an emergency from developing. The first necessity is that you trim for level flight. If, after a time, you find you're using a little back pressure to maintain altitude, or you observe that you have lost some altitude, consider the possibility that the descent may be due to a loss of power. Before adding more power, pull on carburetor heat.

The standard sequence of events if you have ice will be to see first a drop in RPM, then a rise, then another rise after you get rid of the heat. The final reading may look almost exactly the same as the original reading, but that first rise suggests that there was a little ice. Sometimes application of carburetor heat will cause roughness, and I once saw an immediate rise of 500 RPM. If you're in doubt about whether or not you're getting ice, it's best to just check it often.

If your engine quits when you've just made some change, undo whatever you did. Maybe you pulled the mixture knob instead of the carburetor heat. Or switched to an empty tank. Or maybe the engine-driven fuel pump isn't working and you need to go back to the electric one. Undoing what you've just done goes for roughness too, though of course roughness may occur when clearing out carburetor ice. And remember that you may be able to keep an engine running on one magneto when it won't run on both, or you may be able to find a throttle setting which will eliminate roughness, or you may be able to get just enough power to clear an obstacle by using the primer. With care and good judgment you should be able to fly for thousands of hours without cause for so much as one white hair.

Recovering from Inadvertent Entry into Instrument Conditions

Your instrument training is not designed to make you a competent instrument pilot. It should give you the ability to get out of instrument conditions if you encounter them unexpectedly. I feel sure that you'll acquire the necessary ability; the question is, will you use it when you need it? If you become involved in marginal conditions, as might happen through a combination of poor supervision and your own lack of experience, the most important thing to do is *go immediately and completely on the gauges*—start flying by the instruments and only by the instruments, even if you can still see patches of clear ground. Not infrequently you'll read about an accident in which an instrument-rated pilot, flying in marginal weather conditions, and not on an IFR flight plan, has come to grief. It seems puzzling that this sometimes happens even to current and highly experienced instrument-rated pilots. I've wondered if it could happen to me. I don't think so, but then how does it happen to others even more experienced than I am, as well as to beginners. I have speculated on the cause and I think I know what it is.

Let's suppose an experienced instrument pilot takes off in minimum VFR weather. He doesn't file an IFR flight plan because 1) his flight will then take longer, 2) he knows he can file in the air if the weather requires it, and, sometimes, 3) the airplane isn't equipped as he would choose if he were planning an instrument flight. Unexpectedly, he finds himself in full IFR conditions—probably when he's climbing out from an uncontrolled field, though sometimes it seems to happen on an approach. Because he wants to avoid the nuisance of the IFR system, he doesn't immediately start flying by the instruments. Instead, he keeps looking outside to see if he can see the ground, or if it's at night, lights on the ground. While he's doing that—*and not flying the instruments as he's perfectly capable of doing*—he loses control of heading, or altitude, or both and it's all over.

It was something that happened before I was even licensed that first made me realize that even a competent instrument pilot may lose control of an airplane simply because he isn't doing what he's been trained to do. My husband already had his private certificate and was working on an instrument rating. I often went along in the back seat of the Cessna 172 when he took instruction under the hood, and I had seen him fly by the instruments for an hour or more with no difficulty —not only straight and level, but approaches too.

On this occasion he, I, and the three children had been visiting friends on Cape Cod. We had already overstayed our invitation by three days because of bad weather—fog and rain. Finally there was sun. When we took off, it looked pretty hazy and there were scattered clouds where none had been reported, but we got up above the clouds in the sunshine and it seemed OK. Then we started across the water toward the mainland, and suddenly we could see no horizon at all. There was still sun, but we were flying into wispy clouds, and the horizon was lost in haze. We could still see the water straight below us, and my husband began to look, as I did too, down and around for references. Then I noticed that the gyro was starting to turn, and to turn pretty fast. We were starting to spiral and hadn't known it. I said quietly, "You fly the instruments. I'll look outside." He used the instruments then, leveled out, turned, descended and we got safely into New Bedford where we landed to check the weather again. It was deteriorating, so we left the airplane there and got safely home to Pennsylvania (driving through an unforecast squall line) with a rental car we could ill afford.

The important point to that story is that if I hadn't reminded him to go on the gauges when I did, he would certainly have lost control of the airplane. Given the instrument training he had had, and the high altitude we started at, he probably would have regained control; but how much easier, safer, and less scary it was just to maintain it. Believe it or not, if the ground is there to be seen, and if it is clear enough for you to fly by visual references, you will know it even when your eyes are glued to the instrument panel.

Suddenly the Ground Disappears

People do fly straight into clouds without any idea that it's about to happen. It doesn't happen unless the pilot has made some mistakes already—like flying between cloud layers, or in heavy precipitation, or in bad haze—but it does happen. If you should do it, don't start looking around to see if you're really in a cloud. Look immediately at your artificial horizon and make sure your attitude is level. Then check your heading and altitude. Then decide what you want to do, and do it. Take it easy. Don't be in a rush. If you're going to descend, decide on the minimum altitude you'll descend to, and what you'll do if you're still in the soup. If you're going to turn around, don't even start the turn until you're sure what your new heading ought to be.

Another way that people get into clouds unexpectedly is to climb into them. This can even happen climbing out from the airport on a hazy or gray morning. You may not realize that there are low scattered to broken clouds, perhaps even below pattern altitude. And incidentally, such conditions can be very spotty, so even a good check of weather at nearby reporting stations may not give you any warning. Anyway, the remedy is the same: *Go immediately and completely on the instruments.* Check your attitude, and your heading and altitude. The airplane should already be trimmed for the climb. If you aren't 500 feet above the ground yet, or if you have mountains, towers, etc., nearby, then keep on climbing. Just watch the heading and wing attitude closely. Remember the airplane does have a tendency to turn to the left in a climb and will probably be needing some right rudder.

If you are already above 500 feet above the ground and there aren't any obstacles in the area to worry about, level the nose and the wings, hold the heading, and get yourself trimmed for level flight. Don't look outside, even if you catch glimpses of the ground out of the corner of your eye. Once you have the airplane set up for stable flight, decide what's the best thing to do next. Your choice depends on the terrain and on how high above the highest nearby obstacle you were when you ran into the clouds. Another point to consider is the possibility of other traffic, maybe IFR traffic, up there with you. If you have at least 500 feet between the base of the clouds and the ground, and if the ground is flat, you may want to think about letting down gradually, straight ahead, until you have outside references and visibility again. Then you can return to the airport using ground references. But even landmarks you know well are hard to use from 500 feet, so don't try this unless you know the area very very well—better than you're likely to know it as a student pilot.

While you're letting down, watch the altimeter and the directional gyro and the artificial horizon and don't even glance outside until you know, *without looking out,* that the ground is clear enough all around to give you a good outside reference for visual flight. Do set a minimum altitude for yourself. The clouds that had a base at 600 feet may be down to 400 feet by now (or, more happily, up to 800 feet), so you can't just descend until you break out without being concerned about the altimeter. If you don't break out—completely out—by the time you're down to the altitude you've set as a minimum, level out, set up a climb (to above the cloud layer, if that's reasonable), and fall back on your alternate plan. That would be to contact the nearest FSS or radar facility for assistance to an area of higher ceilings or to an airport where you can get a radar approach. You may be inclined to object that you ought to call the controller concerned with the area before you climb. True, technically you ought, and if you have the airplane well under control, by all means do so. But your first job always is to fly the airplane, so don't let concern about communication distract you from that most important job. Think of the four C's: CONTROL the airplane, COMMUNICATE with the controller, CONFESS your predicament, COMPLY with instructions.

If you remember only one thing out of all that's in this book, make that one thing the rule, *Go immediately and completely on the gauges.* That, in itself, is worth the price of the book. I have to admit that I think it's pretty funny that

I should be saying this. After all, my general message is that the joy of flying lies in flying by all your senses, rather than by the instruments, and in developing and using self-discipline and good judgment so you never get into a spot where you'll need to go on your gauges unexpectedly. But we aren't gods and we won't even grow up to be gods when we get older, so we do keep on making mistakes. The trick is to live through them and learn from them. This advice can help you do that.

Night Flying

Somewhere along the way your instructor will schedule the three hours of night work. It's nice to have at least part of it before you go cross-country solo, but you'll probably be finishing it up near the end of your training. This time should include a night cross-country, as well as takeoffs and landings. It would be a good idea to see what it's like on a bright, moonlit night as well as on a really dark night. It used to be that lots of pilots flew for hundreds of hours without ever flying at night. Most of the airplanes didn't have electrical systems and lighted fields were few and far between. That's not true today, and once you've tried night flying you may want to do a lot of it. It has a lot of advantages. There's less wind, and there's smoother air. If it's hazy, you may be able to see landmarks from farther away than you can in daylight—lights come through night haze much better than cities through day haze.

Night cross-country feels more comfortable with radio navigation, at least as an adjunct to dead reckoning and pilotage. Pilotage at night is quite practical in a lot of areas, provided you keep careful track of time and distance, but the landmarks mostly are different and there are fewer of them. And the charts don't seem to have been designed for night flying; it's hard to read them by red cockpit lighting. Go ahead and use the white map light. Yes, it does interfere with your night vision, but what you're looking at on the ground is lights anyway.

Those yellow cities really come into their own at night; you can often match shapes exactly. But water becomes almost useless as a reference. You may be able to spot large rivers and the bridges across them, but even that isn't as easy as you'd think. Far better is the pattern of highways which is more striking at night and can be seen from farther away than in the daytime. And there are two things we don't make much use of in daylight that are very valuable at night—towers, which are lighted with red lights and may also have strobe lights on them, and lighted airports. It is not unusual to be able to see the beacon of a large airport from over thirty miles away, even during periods when prevailing daytime visibilities are ten or twelve miles or less. And lights of large cities are also visible for many miles, sometimes so many miles that it's confusing.

Night flying does have some drawbacks. First, unless you are a natural night owl, you are probably not as sharp as you are earlier in the day. Second, you can't see what's happening to the weather, so if there is unforecast deterioration—especially with regard to ceilings—it can come as more of a surprise at night. Third, you may have trouble judging which way to land at the smaller fields, where there may be a lighted runway but no lighted wind indicator. One thing you can do in this case is to call a nearby controlled field asking for local wind, but remember the wind may be different where you are. There is a way to determine before it's too late whether you are coming in to land downwind. While on base, line up two lights ahead of you and check your drift. Drift is quite easy to see at night using lights, and you should be able to figure out whether you're drifting toward the field, which would mean you were about to land the wrong way. You may also have trouble figuring out where to taxi on a big field. Those blue lights can be hard to make sense of after you're down on the ground. Finally, you really have a problem if you have to make a forced landing. That's one reason for including pilotage in your methods when you go cross-country at night: You want to know where the nearest airport is all the time.

For some people the principal reason for not flying at night is a phenomenon called "automatic rough." This is what an aircraft engine goes into when you are flying over water or at night. Some people think it's all psychological, brought on by the knowledge that a forced landing will be pretty chancy. It seems to me that there's more to it than this. The pilot's nervous condition is the main thing, but I think it's partly an effect of flying in very smooth air with not much to look at so you listen more than usual. Such smooth air is uncommon at the altitudes beginners use—except over water or at night. In very smooth air, natural variations in the sound or feel of the engine's vibrations become noticeable as they would not normally be. At night or over water these variations—magnified by the pilot because of his nervous state of mind—become engine roughness, from which we have "automatic rough."

The Minimum Requirements

When you complete all the minimum time requirements—twenty hours solo, ten hours solo cross-country, three hours night—and realize that you still aren't ready for the flight test, you may begin to feel that you just don't have what it takes. Remember that I said the forty-hour minimum was set a long time ago, before radio, before instruments, before night flying? I'd like to say a little more about that pressure to get through in the least possible time. Even among the 85 percent or so who do pass the flight test the first time, most have well over forty hours and many pass only by the skin of their teeth. A new minimum of seventy-five hours is what I'd like to see, including in the required experience at least two models of airplanes, a variety of airports, and a really long, overnight cross-country. And I'd favor an intermediate certificate for local-only passenger-carrying which would carry the present student certificate requirement for a flight instructor endorsement every ninety days. After a seventy-five hour course almost everyone could pass the test the first time around, and equally important, preparation for real cross-country flights—across hundreds of miles and many hours or days of time—would be more realistic. The FAA objects that they couldn't police a local-only certificate—not a good excuse since they can't police student certificates now.

The emphasis on minimum time wouldn't matter so much if everyone kept on flying regularly and continued to practice and improve. But most newly licensed pilots fly less often than they did when they were taking lessons. When they do fly, they fly with friends or family for fun or to go somewhere. They certainly aren't going to be practicing stalls, for example, on such flights, and they usually don't work very hard at their flying either. All too many pilots fly better on the flight test than they ever will again.

Of course, you are eager to introduce your friends and loved ones to the wonders of flight in a light plane. And it is ego-building to demonstrate your ability as pilot-in-command to a plane-load of admiring groundlings. They'll never know the difference if you wander from side to side and up and down while you're flying in the air; only the landing really counts with nonpilots.

But think. Do you really want to trust your friends and family, let alone yourself, to the flying of a just-barely-passing, D-minus pilot? Of course not! Remember that you do not become an A-plus pilot by meeting the minimum standards in the minimum time, but by meeting high standards no matter how long it may take. Plan to fly five more hours of good hard practice after the instructor has said you're OK for the recommendation ride. Those five hours represent all that makes a pilot a good pilot: prudence, effort toward proficiency, and above all, self-discipline in the face of strong temptation. They'll do more for your ego than any other five hours in your training, and you'll deserve every bit of the glow of pride and confidence you'll feel when you complete them.

14. You Get the "License to Learn"

The Flight Test

The recommendation ride comes before the flight test, but I'm going to talk about the flight test first because your instructor will gear your recommendation ride to the flight test he knows you'll be taking, not the other way around. Your flight test will be given in three parts: an oral examination on the ground first; a demonstration of your cross-country flying in the air; and a demonstration of your flying skills including some instrument work. The oral part will cover all the areas that you had trouble with on the written test (as determined from the Written Test Results form). The examiner will also discuss and ask questions about things he feels are particularly important based on his own experience. Most instructors have a list of the questions frequently asked on oral examinations, and will go over those as well as the questions you missed on the written test and will make sure you thoroughly understand them.

A thorough oral examination may take an hour or even more. In addition to answering questions, you will present your planning for a two-hour cross-country flight. Some examiners will tell you what your destination will be when you call to make the appointment for the test. Others feel you should be able to do an adequate job of planning in twenty to thirty minutes and will expect you to demonstrate this ability after you arrive for the test. Your instructor will know how the examiner usually does it, and if you need to be ready to do the job within limited time he'll help you figure out what's most important and what can be left out or done in the air. You will probably have to demonstrate a pre-flight inspection and it won't do a thing for your confidence if the tire pressure is low, or the windows are dirty, screws are missing, the prop is nicked, etc. So take some extra time beforehand to be sure the airplane is in tip-top condition.

The first part of the flight portion of the test will be a demonstration of cross-country flying as you start on the flight you have planned. You will usually fly no more than twenty-five or thirty miles—long enough to establish a good heading and to make a couple of ground speed checks and estimate arrival time at your destination. Most examiners will give you plenty of time to find and to correct any errors you may make. In fact, it's usually true that the longer the flight test the worse the job the applicant has done. But the average length of a flight test varies enormously from examiner to examiner, so don't start comparing flight test lengths among the pilots you know.

After he's sure that you have established a good course and have a reasonable estimate for your destination, the examiner will probably ask you to figure a heading and an arrival estimate for some point off your course, and to fly there as if you had met a line of thunderstorms, for instance. You may be asked to make a landing at this unexpected point. And it's likely to be a field with some special problem —a right-hand pattern or an approach over obstacles—so be careful in your check of the field from the air. Then you'll take off again and gradually work back to the examiner's home base doing the instrument work and the airwork on the way. The time spent on the test depends not only on the examiner, and on your skill and efficiency, but also on weather, traffic, and other things which are unpredictable. Probably the average total time for the private test is two and a half to three hours, of which about an hour and a half will be the flight portion.

The Recommendation Ride

Your recommendation ride with your own instructor should be run as a simulated flight test. There should be virtually no dual involved. This may seem wasteful of the instructor's time, and most instructors do give active dual during the recommendation ride. But this tends to result in reduced confidence on the actual flight test, and possibly in marginal performance as well. Ideally, your performance on the recommendation should be up to and better than the minimum standards necessary to pass the flight test. Your instructor can make notes as you fly and give you a thorough debriefing covering all the points where he feels you can and should make further improvement after you get back on the ground.

If, at some point in the recommendation ride, your performance is so poor that it would fail a flight test, the best way for the instructor to deal with it is to do as an examiner might. Many examiners will tell you immediately, if you fail some portion, that you will be required to return and retake that portion of the test. Then they will ask if you prefer to continue, or would rather cancel the rest of the test. Unless you are so demoralized that you feel completely unable to go on, it is wisest to continue. It may be that some maneuver you have not yet demonstrated is also poor, and this way the examiner can let you know so you can improve that too before you return. Otherwise you may come back a second time and fail a new portion of the test. You will be retested only on those portions which have been failed or omitted on the previous test.

Dealing with Nervousness

No one finds the flight test easy. Even when you are fortunate enough to know the examiner beforehand, he is an unnerving presence when you're taking the test. The effect is called checkitis, and it's characterized by stammering and stuttering, dropping of pencils and losing of charts, and absurd errors like forgetting to untie the tail or remove a wheel chock. No one is immune. I experienced it with humiliating results when I flew the Atlantic with a highly proficient ferry pilot. I didn't check the oil; I landed and taxied off-center on runways and taxiways, and I made an astonishingly bad crosswind approach and landing—so bad that I'm sure my colleague must have wondered how I had accumulated 9000 hours without any crosswind landing accidents. All my errors were typical of primary students—things I had cautioned students about hundreds of times—so I know about checkitis first hand from my own painful experience.

Some examiners do nothing to make it easier on you. This may seem cruel, but there is some logic on their side. They figure if you can't handle this flight test pressure and still fly safely, you probably won't be able to handle the real pressures that will certainly arise from time to time in a long flying career. I don't wholly agree. That is, the applicant is, after all, only a beginner. And I know that even if I try to make it less traumatic, he will still be nervous. But if I try to calm him, and he still isn't able to fly safely, then I am certainly justified in making no allowances for nervousness.

Even if the examiner doesn't try to reduce your nervousness, there are some ways of thinking about it and ways of doing things that will make it a little easier. The first thing is to say to yourself what was told to me by a music teacher when he came to take his flight test with me. He seemed unusually relaxed and I commented on it. He told me it had to do with something he told his music students before recitals, "Go ahead and be nervous. Everybody's nervous. But don't be nervous about being nervous."

The next piece of good advice is, "Fly to please yourself." By now you know (or think you know) how it ought to be done. If you abandon your own judgment and try to do it the way you think the examiner expects you to do it, the flight is likely to turn into a disaster. This was part of my problem in the trans-Atlantic flight. The examiner really wants to know what your flying will be like when you are alone. You may feel that he's in a hurry and getting impatient, but don't respond by failing to look over a field carefully before you begin your approach, or by failing to clear yourself before you demonstrate a stall. He can only assume that's the way you always do it.

Communicate with the examiner while you're flying. Let him know what you're thinking. Talk out loud about what you're doing. If he asks you to do something, using an unfamiliar term, and you aren't quite sure what he means, don't go ahead and guess. If you guess and do the wrong thing, he may think you don't know how to do what he meant to ask you for. If he asks for something you think is wrong to do, question it. You are the pilot-in-command. Most examiners don't set intentional traps. Nor did I, but I did ask a student once for a stall when he was only at 1200 feet. It happened that he had lost over 1000 feet during the slow flight demonstration he had just finished. I usually follow slow flight with the stalls, so I did as usual, in part to see whether he knew that he had lost altitude. He didn't, and he remained a student a bit longer.

If you do a maneuver that you yourself know doesn't quite meet acceptable standards, and if you are sure you can do it better, tell the examiner you'd like to try it again. He will probably allow that. Above all, if you really make a mess of something, don't try to excuse it away by blaming it on the weather (you had the option of canceling the test), nervousness (he knows about that already), or your instructor (he's heard that before, and he won't buy it). Accept the responsibility yourself. It shows better judgment, and judgment is the most important thing the examiner is trying to assess. I knew an instructor who actually advised his students to set up the first short-field approach on the flight test a little too high. Then, in going around, the applicant would be able to demonstrate his good judgment to the examiner.

Flight Testing Can't Be Entirely Fair

It is very depressing to fail a flight test. Of course, you won't fail if you do a really good job. But a performance that would pass with one examiner may not be good enough with another. Since you, of course, are planning to be well above average, that won't bother you; but I'd like to explain why it's so. The FAA has found that there is a certain percentage of failures among applicants who come to the FAA inspectors for testing. So they put a lot of pressure on designated examiners to maintain a proportion of failures, on a yearly basis, which is at least close to that percentage. The result is that if you fly with an examiner who tests a lot of applicants whose skills are marginal, your fair performance can be passed easily. But if you fly with an examiner who tests mostly well-trained applicants, the same passable performance may put you on the borderline or below it. It's not fair, but that's the way it is. I don't mean to sound as if I think too many applicants are failed. I don't; in fact, I think too many are passed. But I would like to see more consistency.

There are other things that make flight tests less than entirely equal across the country. Prevailing visibility and the type of terrain and landmarks have a lot to do with the ease of the cross-country portion. The places where a visibility of sixty miles requires an explanation of what is "restricting" vision offer a very different situation from those where sixty-mile visibility might occur a couple of times a year. Out in Oklahoma, where I learned to fly the helicopter, they think the visibility is lousy if they can't see the mountains seventy-five miles away. Along the east coast we think "better than fifteen" is the same thing as "unlimited." Well, life isn't fair, and you do the best you can with what you've got.

Dual for the Licensed Pilot

I guess all examiners have their favorite cautionary tales and words of wisdom with which to speed the parting brand-new pilot. But there seem to be some things that aren't mentioned often enough. There are pitfalls waiting especially for the unwary newly licensed pilot. One of them is a four-place airplane loaded to maximum gross weight with baggage and people. You may have all your flying time in this airplane, or you may get checked out in it after you are licensed. In either case, you will be used to its performance with only one or two people in it. If you aren't well warned, you'll be disbelieving when you see the difference fully loaded. All the charts in the manual won't really prepare you for the long takeoff roll and the flat climb. Add the effect of high density-altitude, especially unfamiliar high density-altitude, and you may be in real trouble.

Here's what can happen. You've been flying in fall, winter, and spring, either alone or with one or two other people and no luggage. Now summer comes and you set off on a vacation flight with a fully loaded plane. You "know" what the airplane performance is—you've seen what it will do with regard to runway length, climb angle, glide angle. But you've never seen or imagined how much the airplane performance is reduced by the full load combined with heat, especially humid heat. The numbers in the manual just can't convey the picture.

Your takeoff roll will go on . . . and on. Even if you don't succumb to the temptation to pull the airplane off a little too soon, the flat climb after liftoff may trick you into raising the nose higher to get better performance. Too much reliance on attitude can be a trap in this situation, because the attitude that looks familiar and right isn't right even for a short-field maximum performance climb, and may even produce a stall.

To head off this problem there are two things you can do. First, get a couple of friends to come along while you get some additional dual on short-field takeoffs and landings. Second, simulate high density performance by using less than full power for takeoff. Don't do this solo and don't do it if your instructor disapproves, and don't overdo it—your engine doesn't get proper cooling when you treat it this way. But it will give you something like the actual climb angle you'd get with a full load under adverse conditions of temperature, pressure, and humidity and will let you see how the pitch angle is affected. It's worth doing a couple of times at least, just to see how long the takeoff roll can be, and just how flat the climb is.

It is not only summer that holds hazards peculiar to the

season. Many pilots are licensed today with experience of flying in only one or two seasons. The weather and other conditions peculiar to the seasons you aren't familiar with probably haven't even been discussed. Give this some real thought. Blustery winter winds are a lot different from summer ones. Navigation is different when there's snow on the ground and ice on the water. Ground fog has its season, and so does thick haze. It's not a sign of weakness to sign up for more dual after you're licensed, and it might be a very good idea during those first hot days of summer, or after the first snowfall.

Theoretically, you are qualified to make unlimited cross-country flights once you are licensed. But both terrain and general weather conditions, especially visibility, differ widely in different parts of the country. Mountain flying has its special hazards and deserves some special dual with an appropriately experienced instructor. Be prepared to find that navigation and planning and anticipating weather changes are easier in some parts of the country than in others.

Pilot Error

Of all the hazards you can possibly face by far the most dangerous is in the workings of your own mind and emotions. Only about 10 percent of all accidents can be blamed, even in part, on mechanical failures of any kind, and some of these too can be chalked up to pilot error because there were clear indications of problems before the flight began.

Pilot error can be classified into four distinct categories: the error which is a momentary but crucial lapse; the error which is design-induced; the error which is simply lack of sufficient skill in controlling the airplane; and the error which is poor judgment.

The momentary lapse is easy to describe, but harder to

explain. There's the fellow who took off with no rudder on the airplane, for instance. Less spectacular but equally absurd is the one who had the fuel selector set to the empty tank, or who never put the oil cap back on, or who misread the windsock, or who forgot to change tanks. These are obvious, clear-cut mistakes. The accident reports usually don't offer enough information to suggest how such mistakes could come about. It looks like just not paying attention, and there could be a lot of reasons for that—being upset, worried, angry, or simply chatting with friends or passengers. Sometimes this kind of error is immediately disastrous. Sometimes it is only one of a number of errors which finally culminate in disaster. If the pilot is lucky, the disaster he's earned doesn't occur; then he can write an article about what he's learned and you can benefit from it.

The design-induced error is a factor for everyone, including the pilot with a lot of flight time. Our tendency to steer with the wheel is an example of that kind of error. There are other examples and some of them have been blamed for fatal accidents, even on airliners. Mostly these are due to changes in instrument presentations, changes which can lead the experienced pilot into misreading the instrument when under pressure. The new-style turn indicator, for example, looks as if it is indicating pitch and in at least one case was read that way for long enough to put an experienced pilot in serious trouble in a thunderstorm. He recovered from his dive (in a DC-3) without structural failure, and lived to tell the tale. But the same thing may well have been a contributing factor in accidents where no one lived to tell us what happened.

Error induced by design can be anticipated, and you can avoid serious consequences by using good judgment. If you fly an unfamiliar airplane with a different instrument set-up or different controls, be especially careful not to fly when you're tired, or likely to be under pressure. "Design-induced" can also refer to things like carburetor heat knobs that can easily be confused with mixture control knobs, fuel selector valve handles that can be moved by a careless foot, or a control that feels familiar but actually works in a direction opposite to the one that it feels like. If you check out in a faster airplane in order to use it for a special trip, be particularly careful about this kind of thing, and set your personal weather minimums a little higher for a while.

The third type of error, simple lack of skill, most often results in serious accidents when the deficiency is in the area of speed control—stalls and spins. I've said enough about speed control already, so I'll just point out that these problems usually become critical only after the pilot has already used poor judgment—following someone too closely in the pattern or waiting too long to go around.

Judgment

Judgment is the big element in pilot error. Good judgment can keep you from having the momentary lapse—careful, deliberate use of checklists, adherence to the basic rules of good sense, would prevent almost all of these. Good judgment can also reduce the effect of poor design or of design changes. Good judgment can keep you from exceeding your skill level. And I'll repeat myself and say again that most of all judgment depends on knowing yourself, being honest with yourself, and accepting your own limitations. Be aware of your current state of mind. If you know you're upset, don't fly. Or if you do fly, be doubly careful about doing everything precisely as it should be done.

A lot of us suffer from a drive to be supermen and superwomen. We feel that we *ought* to be able to rise above mental or physical troubles and fly safely and well through sheer strength of will. Well, maybe we can; and there are some people who would rather die trying than "give up" while there's still a chance. That's fine, if it's a matter of life or death, but is it really necessary to put it to the test in your own flying career? The only life or death involved is yours (and those of your passengers). You aren't delivering life-saving serum. Wouldn't it be better to stay home wishing you were flying than to be flying wishing you had stayed home?

Very few accidents can be blamed on anything as simple as taking off without a rudder. Most accidents happen only at the end of a long series of errors. And it is surprising how often we can see in the clear vision of hindsight that the first mistake was the decision to take off in the first place. The basic reason for the accident is that the flight should never have been made.

So be very careful about your decisions to fly. If you tend to be an optimist, be especially careful. Try to look without prejudice at the things that you're taking into account as you make your decision. If you find that some of the factors have nothing to do with flying as such, then weight the factors on the side of staying on the ground extra heavily.

Factors that have nothing to do with flying are things such as having promised a ride to a friend, or having planned a trip for weeks in advance, or even being on the way home and only an hour's flight away with night coming on and your family tired and hungry. Look at the following hypothetical situations and think about what you would be likely to do.

One: You've promised your boss a ride. He's never been up in a light plane, and he calls you full of enthusiasm, saying, "It's a great day. I can hardly wait. What time shall I be at the airport?" You know that the clear blue sky after last night's cold front means the wind is soon going to be howling across the runway—blowing a lot harder than your usual limit. But you don't like to tell your boss that there's anything you can't handle.

Two: You've been planning a weekend trip for weeks. It's the first one with your wife. You're going farther than you've ever been, into the mountains. The day dawns hot and humid. Your wife talks eagerly about how nice and cool it will be in the mountains. You check weather. There is haze, three to five miles visibility, a few leftover thunderstorms, and showers and thunderstorms are forecast to be possible all weekend. You've always avoided cross-country flying in haze.

Three: You're on the way home from a week's vacation with the family, broke but happy. Forty miles from home you run into showers, static on the radio, marginal visibility, and it's looking pretty dark ahead. You know there's a nice airport five miles off to your left with a motel right there, a restaurant, and a Weather Bureau office too. But it won't be cheap, and home is only half an hour away. You know there's a front forecast with thunderstorms and high winds, but it isn't due till after you'd be home.

That last one is especially insidious because, for one thing, you're already up in the air, following a particular plan; and for some reason, especially when we're tired, we humans have a tendency to carry on without reassessing our situation. This one is a classic set-up for a fatal accident. You are tired, and you're anxious to get home. You're worried about the expense. You may be worried about what your wife is going to say the next time you want to go on a flying vacation if you can't stay on schedule with this one. Home is so near. You're familiar with the ground and landmarks and you know you can't get lost. You tend to feel safe when you see familiar territory, even when the weather is stuff you would never have taken off into.

Think hard about those situations. Relate them to accidents you read about. Ponder the way people's minds work, and look very critically at the decisions you make.

Safe Flying

I've put a lot of my feeling about what makes flying safe into these comments. I want to add a little more. I've been told by some of my students, even years later, that they still hear with an inner ear some of the things I've said to them. A simple rule or catch phrase may be enough to save your life, if it comes into your head at the right moment. Here are a few of my favorites.

When I started flying I realized that I wanted to be accepted as the equal of the men around the field. I was afraid that desire might trap me into doing things I shouldn't do, just to prove I wasn't a weak and frightened female. So I made up a catch phrase to help me resist that temptation: "It's better to let them say, 'She could have made it,' a thousand times, than to let them say, 'She shouldn't have tried,' once." There have been times when I've needed that little extra nudge on the side of prudence.

Another very good one is, "When in doubt, don't." If you are honest with yourself, you will acknowledge the doubt you feel. Here I'm really talking about what one of my students called "a small fear." Edmund Burke said, "Early, provident fear is the Mother of Safety." If you don't fly when in doubt, you will fly well inside the limits of your skill most of the time, and you'll have a margin of competence to handle the unexpected. It's a good rule to use in everyday life too. But it depends first on the ability to be honest with yourself—on being alert to inner doubts and acknowledging small fears.

Finally, I like "Don't do it today, if you wouldn't do it every day." That covers things like taking off with low fuel because there happens to be a line at the gas pump, or with a magneto that's rough and won't clear up, because you aren't at your home field. It's true that most of the time these things won't hurt you, but once in a while a small error in judgment turns out to be the first in a series of mistakes that ends in an accident. We could be talking life and death, so keep the odds as high as you can in your favor.

Some Last Words

Perhaps you're wondering where the joy is in all this talk about responsibility, self-discipline, and such. I think joy must rest on a base of confidence and self-esteem. I gained confidence and self-esteem as I learned to trust my own judgment, to fulfill my responsibilities, to exercise self-discipline. Flying gives more opportunity for growth as a person than anything I know. And it doesn't have to be stressful or dangerous, though it is sometimes allowed to be.

I've talked to a lot of nonaviation groups, especially about the helicopter flight across the country. One question I'm often asked is, "Did you ever think you might not make it?" Well, the answer is, "No, definitely not. Such a thing never occurred to me." If I were asked to drive a car across the country—yes, that thought might cross my mind. But the hazards that are imposed by the aberrations of the other guy are almost nonexistent in the air. On the road I can only be as safe as the guy coming down the street allows me to be. But in the air I am as safe as I choose to be. I personally choose to be safe—very safe—and I am as confident about a successful flight as I am when I put my feet on the floor in the morning that I'm going to make it to the kitchen for a cup of coffee.

It is hard for people who don't fly to imagine what it is like. Groundlings tend to assume that flying is like driving, only in the sky. It's not like driving, though it is hard to explain just how it is different. Some people even think it must be boring. Pilots themselves sometimes say that flying is "hours of boredom punctuated by moments of sheer terror." I have never been bored in the air. I've been asked by nonpilots what I think about when I'm flying. I think about the job I'm doing—the heading, the time, the gas. I think about other traffic. I think about the ground beneath. Sometimes I'm thinking, "Where is that water tower that I ought to be able to see," or "Why isn't that big lake on the chart?"

But if all is going well, I have time to think in broader terms about the pattern that man imposes on the ground: endless rectangles of various shades of green and brown; roads and the strips of buildings on each side of them; miles and miles of cleared leveled ground covered with houses and stores and streets, but with never a trace of running water; farther out from the city (itself almost always built on a river or bay or lake), the innumerable twining lines of bushes or trees that show me where the water runs. I notice how much more open ground or untouched forest exists than a man who is confined to highways could ever know. I enjoy the passage of the seasons in the changing colors on the ground and the changing weather in the sky.

If the ground is out of sight, or so familiar that I hardly see it, there is the sky and the weather to think about: the flow of air and water over the earth; the cues we get from clouds and wind about what's coming next; the forces upon which man can still impose no pattern. I get a real and present sense of the wonder of life, and I experience delighted astonishment over and over at my presence in the air. I often think, simply, "I feel sorry for people who don't fly. I thank God that I do."

Recommended Books

BACH, RICHARD, *Biplane.* New York, Harper & Row, Publishers, 1966. Avon Books, 1972.

BACH, RICHARD, *Stranger to the Ground.* New York, Harper & Row, Publishers, 1972. Avon Books, 1973.

BACH, RICHARD, *Jonathan Livingston Seagull.* New York, Macmillan Publishing Co., Inc., 1970. Avon Books, 1974.

BACH, RICHARD, *A Gift of Wings.* New York, Delacorte Press/Eleanor Friede, 1974. Dell, 1975.

BACH, RICHARD, *Illusions.* New York, Delacorte Press/Eleanor Friede, 1977.

BERGMAN, JULES, *Anyone Can Fly.* New York, Doubleday & Company, Inc., 1964. Revised edition, 1977.

BERNHEIM, MOLLY, *A Sky of My Own.* New York, Macmillan Publishing Co., Inc., 1974.

BUCK, ROBERT N., *Weather Flying.* New York, Macmillan Publishing Co., Inc., 1970. Revised edition, 1978.

BUCK, ROBERT N., *Flying Know-How.* New York, Delacorte Press/Eleanor Friede, 1975.

COLLINS, RICHARD L., *Flying Safely.* New York, Delacorte Press/Eleanor Friede, 1977.

EARHART, AMELIA, *The Fun of It: Random Records of My Own*

Flying and of Women in Aviation. Detroit, Michigan, Gale Research Company, 1975 (reprint of 1932 edition).

EDITORS OF *Flying* Magazine, *I Learned About Flying From That!* New York, Delacorte Press/Eleanor Friede, 1976.

FEDERAL AVIATION ADMINISTRATION, *The Private Pilot's Handbook of Aeronautical Knowledge.* Washington, D.C., U.S. Printing Office, 1971.

FEDERAL AVIATION ADMINISTRATION, *Terrain Flying.* Washington, D.C., U.S. Printing Office, 1967.

FEDERAL AVIATION ADMINISTRATION, *Aviation Weather.* Washington, D.C., U.S. Printing Office, 1975.

GANN, ERNEST, *Fate Is the Hunter.* New York, Simon & Schuster, Inc., 1961.

HELMERICKS, HARMON, *Last of the Bush Pilots.* New York, Alfred A. Knopf, Inc., 1969.

KERSHNER, WILLIAM, *The Student Pilot's Flight Manual.* Ames, Iowa, Iowa State University Press, 4th edition.

LANGEWIESCHE, WOLFGANG, *Stick and Rudder.* New York, McGraw-Hill Book Company, 1944.

SAINT-EXUPERY, ANTOINE DE, *Airman's Odyssey.* New York, Harcourt Brace, Inc., 1943.

SAINT-EXUPERY, ANTOINE DE, *Wind, Sand and Stars.* New York, Harcourt Brace Jovanovich, Inc., 1967.

SAINT-EXUPERY, ANTOINE DE, *Night Flight.* New York, Harcourt Brace Jovanovich, Inc., 1974.

SCOTT, SHEILA, *Barefoot in the Sky.* New York, Macmillan Publishing Co., Inc., 1974.

SMITH, D. C., *By the Seat of My Pants.* Toronto, Canada, Little Brown & Co., Ltd., 1961.

SMITH, FRANK K., *Week-end Pilot.* New York, Random House, Inc., 1974.

TAYLOR, RICHARD L., *Instrument Flying.* New York, Macmillan Publishing Co., Inc., 1972. Revised, 1978.

TAYLOR, RICHARD L., *Fair-Weather Flying.* New York, Macmillan Publishing Co., Inc., 1974.

TAYLOR, RICHARD L., *Understanding Flying.* New York, Delacorte Press/Eleanor Friede, 1977.

Index